CONTENTS

PREFACE

What do an Olympic athlete, your favorite music artist, and Albert Einstein have in common? They all became experts in their fields through practice. To understand physics and to do well in your course, you must practice. When you learned to walk, ride a bike, and drive a car; you had to practice to master those skills. It would be silly to think you can learn physics by listening to lectures and skimming the book. This study guide is designed to help you practice and to build a deep understanding of physics.

Expert problem solvers in physics follow a systematic approach in their problem solving. Elite athletes also follow a systematic approach in their training to reach the upper level of their sport. You should also follow a systematic approach in your physics course to fully develop your skills. To encourage you in building good problem-solving skills, this study guide follows a systematic problem-solving procedure throughout–the **Identify**, **Set Up**, **Execute**, and **Evaluate** procedure developed in the textbook.

In the **Identify** phase of the problem, you should identify the relevant concepts. Decide which physics concepts can be used to solve the problem. Identify the target variable in the problem, and keep this target variable in mind as you solve the problem. Don't think you can save time by skipping this step and jumping right into an equation search. You need to plan a strategy for solving the problem: Decide what you know, where you are going, and how to proceed to the solution.

In the **Set Up** phase of the problem, you should select the equations you will use to solve the problem and how to use them to determine the solution. Make sure you select equations that are appropriate for the physics of the problem, and don't select equations based solely on the variables in the equation. You should sketch each problem to help you visualize the physical situation and guide you to the solution. Rarely do physicists discuss cutting-edge research problems without first sketching their ideas.

When you proceed to **Execute** the solution, work through the solution step-by-step. Identify all of the known and unknown quantities in the equations, making a note of the target variable. Then do the calculations to find the solution, writing down all of your work so you may return and check it later. If you run into a dead end, don't erase your work as you may find it useful in a later phase of the problem. Try another avenue when you get stuck and you will eventually find the solution.

After completing the problem, **Evaluate** your result. Your goal is to learn from the problem, and build your physics intuition. Does the answer make sense? If you were estimating how high an elephant can jump, you'd expect it ought to be less than a meter or two. Consider how this problem compares to the last problem you completed, the example in the text, and the example shown in class. Physicists constantly compare and contrast their new results to previous work as they observe natural phenomena, find patterns, and build principles to connect various phenomena.

Questions and problems chosen for this study guide cover the most critical topics you'll encounter. Working through the guide will better prepare you for homework (including MasteringPhysics) and exams as well as assist in developing a deeper understanding of physics. One way to build confidence is to try working through the questions and problems in the guide for practice, referring to the solutions only when you get stuck. Building confidence before an exam reduces stress during the exam, improving performance. Several 'Try It Yourself' problems are included at the end of each chapter to help build your confidence. Solution checkpoints are included for each of these problems to help you if you get stuck. Summaries, Objectives, Concepts and Equations, and Problem Summaries are ancillary materials that help bring the physics topics of each chapter into coherence. Taking advantage of all of the components in this study guide will help build your problem-solving repertoire.

This study guide is but one of many resources at your disposal when learning physics. Your instructor, class, and textbook are also important resources. But you should also consider who approaches the material from the same level and perspective as yourself–your fellow students. The best untapped resource in a physics class is often other students learning physics for the first time. Discuss physics as a group and confront your questions together, just as many professionals collaborate in the workplace.

We know physics has a reputation for being challenging. While it can be challenging, many students have succeeded in learning physics. Their success was built on a series of small steps, regular practice, and following a systematic approach. Follow their footsteps and you will master physics as they did. You'll also find physics to be a rich and beautiful subject.

Good luck and enjoy learning physics!

Dedicated to Marley, a 45-pound lab mutt that brought a petaton of happiness to our lives.

Laird Kramer
Miami, Florida, 2007

1 Units, Physical Quantities, and Vectors

Summary

Physics is the study of natural phenomena. In physics, we build theories based on observations of nature, and those theories evolve into physical laws. We often seek simplicity: Models are simplified versions of physical phenomena that allow us to gain insight into a physical process. This chapter covers foundational material that we will use throughout our study. We begin with measurements that include units, conversions, precision, significant figures, estimates, orders of magnitude, and scientific notation. We will also examine physical quantities. Scalar quantities, such as temperature, are described by a single number. Vector quantities, such as velocity, require both a magnitude and a direction for a complete description. We will delve deeper into vectors, as they are used throughout physics. We'll also include a summary of critical mathematics skills you will apply throughout your physics career. Techniques developed in this chapter will be used throughout our investigation of physics.

Objectives

After studying this chapter, you will understand

- The process of experimentation and its relation to theory and laws.
- The SI units for length, mass, and time and common metric prefixes.
- How to express results in proper units and how to convert between different sets of units.
- Measurement uncertainties and how significant figures express precision.
- The use and meaning of scalar and vector quantities.
- Various ways to represent vectors, including graphical, component, and unit-vector representations.
- How to add and subtract vectors, both graphically and componentwise.
- How to multiply vectors (the dot product and the cross product).
- The six most critical mathematics techniques you will encounter in physics.

Concepts and Equations

Term	Description
Physical Law	A physical law is a well-established description of a physical phenomenon.
Model	A model is a simplified version of a physical system that focuses on its most important features.
Système International (SI)	The Système International (SI) is the system of units based on metric measures. It established refined definitions of units, including definitions of the second, meter, and kilogram.
Significant Figures	The accuracy of a measurement is indicated by the number of significant figures, or the number of meaningful digits, in a value. In multiplying or dividing, the number of significant figures in the result is no greater than in the factor with the fewest significant figures. In adding or subtracting, the result can have no more decimal places than the term with the fewest decimal places.
Scalar Quantity	A scalar quantity is expressed by a single number. Examples include temperature, mass, length, and time.
Vector Quantity	A vector quantity is expressed by both a magnitude and a direction and is often shown as an arrow in sketches. Vectors are frequently represented as single letters with arrows above them or in boldface type. Common examples include velocity, displacement, and force.
Component of a Vector	The vector $\vec{A}$ lying in the xy plane has components A_x parallel to the x-axis and A_y parallel to the y-axis; A_x and A_y are the x and y component vectors of $\vec{A}$. Vector $\vec{A}$ can be described by unit vectors—vectors that have unity magnitude and that align along a particular axis. The unit vectors $\hat{i}, \hat{j}$, and $\hat{k}$ respectively align along the x-, y-, and z-axes of the rectangular coordinate system. Here, $$\vec{A} = A_x\hat{i} + A_y\hat{j}.$$
Magnitude of a Vector	The magnitude of a vector is the length of the vector. Magnitude is a scalar quantity that is always positive. It has several representations, including $$\text{Magnitude of } \vec{A} = A = |\vec{A}|.$$ The magnitude can be found from the component vectors as $$A = \sqrt{A_x^2 + A_y^2}.$$
Vector Addition and Subtraction	Two vectors, $\vec{A}$ and $\vec{B}$, are added graphically by placing the tail of $\vec{A}$ at the tip of $\vec{B}$: Vector $\vec{B}$ is subtracted from vector $\vec{A}$ by reversing the direction of $\vec{B}$ and then adding it to $\vec{A}$:

Vector addition can also be done with component vectors. For components A_x and A_y of the vector $\vec{A}$ and components B_x and B_y of the vector $\vec{B}$, the components R_x and R_y of the resultant vector $\vec{R}$ are given by

$$R_x = A_x + B_x \text{ and } R_y = A_y + B_y.$$

Scalar Product

The scalar, or dot, product $C = \vec{A} \cdot \vec{B}$ of two vectors $\vec{A}$ and $\vec{B}$ is a scalar quantity. It can be expressed in terms of the magnitudes of $\vec{A}$ and $\vec{B}$ and the angle ϕ between the two vectors—that is,

$$\vec{A} \cdot \vec{B} = AB\cos\phi = |\vec{A}||\vec{B}|\cos\phi,$$

or in terms of the components of $\vec{A}$ and $\vec{B}$—that is,

$$\vec{A} \cdot \vec{B} = A_xB_x + A_yB_y + A_zB_z.$$

The scalar product of two perpendicular vectors is zero.

Vector Product

The vector, or cross, product $\vec{C} = \vec{A} \times \vec{B}$ of two vectors $\vec{A}$ and $\vec{B}$ is a vector quantity. The magnitude of $\vec{A} \times \vec{B}$ depends on the magnitudes of $\vec{A}$ and $\vec{B}$ and the angle ϕ between the two vectors. The direction of $\vec{A} \times \vec{B}$ is perpendicular to the plane in which vectors $\vec{A}$ and $\vec{B}$ lie and is given by the right-hand rule. The magnitude of $\vec{C} = \vec{A} \times \vec{B}$ is

$$C = AB\sin\phi$$

and the components are

$$C_x = A_yB_z - A_zB_y$$

$$C_y = A_zB_x - A_xB_z$$

$$C_z = A_xB_y - A_yB_x.$$

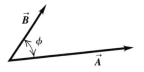

$\vec{A} \times \vec{B}$ is perpendicular to the plane of $\vec{A}$ and $\vec{B}$.

(Magnitude of $\vec{A} \times \vec{B}$) = $AB \sin \phi$

The vector product of two parallel or antiparallel vectors is zero.

Mathematics Review: Top Six Math Skills You Will Need in Introductory Physics

Mathematics is the main language of physics. You will rely on mathematics throughout your study of physics and therefore must become comfortable with mathematical techniques. Here, we present the six most important mathematical techniques you will use throughout your physics career. We strongly encourage you to review these materials thoroughly. When your knowledge of mathematics becomes second nature, your understanding of physics will blossom.

Math 1: Trigonometry

We will use trigonometry throughout physics; problems involving objects tossed into the air, ramps, velocities, and forces all require trigonometry. The basic trigonometric functions relate the lengths of the sides of a right triangle to the inside angle. We define $\sin\theta$, $\cos\theta$, and $\tan\theta$ for the right triangle shown:

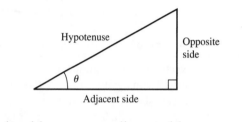

$$\sin\theta = \frac{\text{opposite side}}{\text{hypotenuse}}, \cos\theta = \frac{\text{adjacent side}}{\text{hypotenuse}}, \tan\theta = \frac{\text{opposite side}}{\text{adjacent side}}.$$

Often, the triangle is formed in an xy coordinate system, as is seen in Figure 1.1. Note that the two inside angles complement each other (add to 90°), so two sets of relations can be used:

$$\cos\theta = \frac{x}{r}, \sin\theta = \frac{y}{r}, \text{ and } \tan\theta = \frac{y}{x}$$

$$\cos\phi = \frac{y}{r}, \sin\phi = \frac{x}{r}, \text{ and } \tan\theta = \frac{x}{y}.$$

Figure 1.1 xy coordinate system.

CAUTION   **Watch sines and cosines!** One common mistake is to automatically associate the x-component with cosine and the y-component with sine. As you can see from Figure 1.1, this association does not always hold. One of the most common mistakes encountered in physics is confusing components of sines and cosines. By checking components every time, you avoid this mistake.

We will also need to manipulate trigonometric relations, so we will use common trigonometric identities, including the following:

$$\tan\theta = \frac{\sin\theta}{\cos\theta}$$

$$\sin^2\theta + \cos^2\theta = 1$$

$$\sin 2a = 2\sin a \cos a$$

$$\cos 2a = \cos^2 a - \sin^2 a$$

Other trigonometric identities are in Appendix B.

Math 2: Derivatives

Physics often investigates changes and rates of changes of various quantities. Derivatives provide the instantaneous rate of change of a quantity. Derivatives are therefore the natural choice for finding rates of change in physics. They are especially useful when we have functional definitions of quantities. Speed is the rate of change of position. If we are given speed in terms of a position function, we can easily find the speed by taking the derivative of position. The most common derivatives you will encounter in physics are given in Table 1.

TABLE 1: Common derivatives.

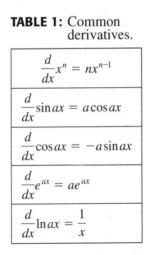

$$\frac{d}{dx}x^n = nx^{n-1}$$

$$\frac{d}{dx}\sin ax = a\cos ax$$

$$\frac{d}{dx}\cos ax = -a\sin ax$$

$$\frac{d}{dx}e^{ax} = ae^{ax}$$

$$\frac{d}{dx}\ln ax = \frac{1}{x}$$

Math 3: Integrals

In physics, we also need to sum various quantities—quantities that are often given in terms of functions. Integrals sum functions and therefore are used to sum physical quantities. For example, one may find the total mass of an object by integrating its density function. The most common integrals you will encounter in physics are given in Table 2. In addition, you may want to review integration by parts and trigonometric substitution for integrals in order to solve the more complicated ones.

TABLE 2: Common integrals.

$$\int x^n dx = \frac{x^{n+1}}{n+1} \qquad (n \neq -1)$$

$$\int \frac{dx}{x} = \ln x$$

$$\int \sin ax \, dx = -\frac{1}{a}\cos ax$$

$$\int \cos ax \, dx = \frac{1}{a}\sin ax$$

$$\int e^{ax} dx = \frac{1}{a}e^{ax}$$

$$\int \frac{dx}{\sqrt{a^2 - x^2}} = \arcsin\frac{x}{a}$$

$$\int \frac{dx}{\sqrt{x^2 + a^2}} = \ln\left(x + \sqrt{x^2 + a^2}\right)$$

$$\int \frac{dx}{x^2 + a^2} = \frac{1}{a}\arctan\frac{x}{a}$$

$$\int \frac{dx}{(x^2 + a^2)^{3/2}} = \frac{1}{a^2}\frac{x}{\sqrt{x^2 + a^2}}$$

$$\int \frac{x dx}{(x^2 + a^2)^{3/2}} = -\frac{1}{\sqrt{x^2 + a^2}}$$

Math 4: Graphs

Graphs are common in many fields, including physics, in which data or information is plotted as a function of time. However, many students do not gain a full appreciation of graphs, and instructors often take graph interpretation for granted. In physics, graphs provide added insight into complex phenomena. We will review important features of graphs to help your physics interpretations, as well as to help your interpretation of graphs wherever you encounter them.

Let's begin by examining the graph in Figure 1.2. Here, position is plotted as a function of time for three different objects. As time increases, the positions of all three objects increase, so the object is moving away from the origin. The slope gives the rate of change of the position—the speed—of the object:

$$\text{slope} = \text{rate of change of position} = \text{speed} = \frac{\Delta(\text{position})}{\Delta(\text{time})}.$$

All three lines are straight lines, indicating that the speed is constant for each object. Both objects A and B start at the same initial position. Object A's line has the greatest slope, so object A moves the fastest or has the greatest speed. Object C starts away from objects A and B and moves away at a slower rate than the other two. Where the lines intersect, objects A and C are at the same position at the same time. After the intersection, object A moves away from object C and so passes object C.

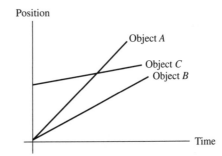

Figure 1.2 Position-versus-time graph.

Moving on to a more interesting case, we see that the slope in Figure 1.3 is not constant; calculus will be necessary to interpret this graph. The rate of change of the position varies, so we will need to consider the instantaneous slope, or the derivative of the position:

$$\text{instantaneous slope} = \lim_{\Delta t \to 0} \frac{\Delta(\text{position})}{\Delta(\text{time})} = \frac{dx}{dt}.$$

Graphically, the instantaneous slope is the tangent to the line in the position-versus-time graph.

If we look at the figure, we see that the object moves away from the initial position, remains at a constant position for a period of time, and then moves toward the initial position. How it moves is found by looking at the slope. The slope of the line increases between point A and point B, indicating that the object begins by moving slowly and then speeding up as it moves away. After point B, the slope decreases until point C, where the slope becomes zero. Thus, the object begins slowing down at point B and stops at point C. At point D, the slope decreases and then becomes constant, indicating that the object moves back toward its initial position. The slope past point D is negative, indicating that the object moves in the direction opposite that of its initial movement. Points E and F show the instantaneous slope at two points on the curve. The slope at point E is greater than the slope at point F, indicating that the object is moving faster at point E.

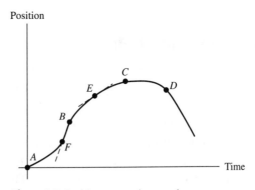

Figure 1.3 Position-versus-time graph.

The speed of another object is plotted in Figure 1.4. From this graph, we can use calculus to determine the distance the object travels in a given time interval. The distance traveled by an object between times t_a and t_b is the area under the curve, or the shaded area shown in the figure. The area under the curve is the speed multiplied by the time, which is the distance traveled, as we will learn in Chapter 2. To find the area, we will need to sum up, or integrate, the speed over time:

$$\text{distance} = \int_{t_a}^{t_b} (speed)\,dt.$$

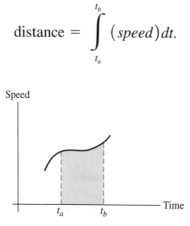

Figure 1.4 Speed-versus-time graph.

We have now seen somewhat how calculus and certain graphs are intertwined. Derivatives and integrals, respectively, give us the slope and the areas under a curve. We will see that switching back and forth between graphs and the mathematics of calculus will help develop our physics intuition and skills. When you're confused about integrals, consider using a graph to clarify your understanding.

Math 5: Solving quadratic equations

We will encounter quadratic equations throughout our physics investigations. We can either factor the equation to obtain the solutions or use the quadratic equation. It is often easier to apply the quadratic equation and not even attempt to factor, as the quadratic equation leads directly to the solutions. For a quadratic equation of the form $ax^2 + bx + c = 0$ with real numbers a, b, and c, the solutions are given by the quadratic equation:

$$x = \frac{-b \pm \sqrt{b^2 - 4ac}}{2a}.$$

Math 6: Solving simultaneous equations

Our goal in solving problems is to determine unknown quantities in equations. When we have multiple unknowns, we will need multiple equations to find solutions. We will need at least as many equations as we have unknowns to solve any problem.

When we are presented with multiple equations, one option is to rewrite one of the equations, solving for one unknown in terms of the other unknown(s), and substitute the result into the other equation(s) to eliminate one variable. For example, if we need to solve for both x and y, given two equations, we first rewrite one equation to solve for x in terms of y. Then we replace the x terms in the second equation with the solution of the first equation, leaving an equation having only y terms.

A second technique comes from linear algebra. We multiply each equation by a factor and then add or subtract the two equations. By choosing the proper factor, we eliminate one variable in the process. For example, with the equations

$$3x + 2y = 3$$
$$4x - 3y = 7$$

we multiply the top equation by 4 and the bottom equation by 3, leaving

$$12x + 8y = 12$$
$$12x - 9y = 21$$

If we now subtract the two equations, the x variable is eliminated. We can also multiply the top equation by 3 and the bottom equation by 2 and add the two equations to eliminate y. The key in this process is to multiply all terms of each equation by the same factor.

Other mathematical topics

Other mathematical relations that we will encounter as we cover the material of this course include the following:

- Circumference, area, surface area, and volume of spheres and cylinders
- Exponentials, logarithms, and their identities
- The binomial theorem
- Power series expansions of algebraic, trigonometric, and exponential functions
- Multivariable calculus, including derivatives and integrals in two and three dimensions.

It is best to review the preceding topics as you encounter them in the course. Appendix B includes a summary of these topics. You may also want to consult your mathematics textbooks or the Internet for more information.

CAUTION **Don't be afraid to review math!** Knowing math will let you focus on physics and save time when you solve problems.

Conceptual Questions

1: Sketch the situation

A man uses a cable to drag a trunk up the loading ramp of a mover's truck. The ramp has a slope angle of 20.0°, and the cable makes an angle of 30.0° with the ramp. Make a sketch of this situation.

Solution

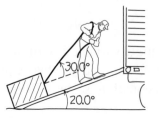

Figure 1.5 Sketch of trunk being dragged up a loading ramp.

IDENTIFY, SET UP, AND EXECUTE: The sketch is shown in Figure 1.5. The ramp makes an angle of 20° with the ground. The trunk is on the ramp and the cable is attached to the trunk. The cable makes an angle of 30° with respect to the ramp, clearly marked. The mover is shown pulling the trunk up the ramp.

EVALUATE: Understanding the physical situation in physics problems is critical for a correction interpretation. You should always draw a diagram (or diagrams) of the physical system you are investigating. Even when a figure is provided, it is often useful to sketch the important aspects. Only after creating a diagram should you proceed to interpret the physics and determine the proper equations to apply.

2: Dimensional analysis practice

Based *only* on consistency of units, which of the following formulas could *not* be correct? In each case, x is distance, v is speed, and t is time.

(a) $\quad t = \sqrt{\dfrac{2x}{9.8 \text{ m/s}^2}}$

(b) $\quad x = vt + (4.9 \text{ m/s}^2)t$

(c) $\quad v = v_0\sin\theta + \dfrac{(9.8 \text{ m/s}^2)x}{v_0\cos\theta}$

(d) $\quad x^2 - \dfrac{2v_0^2\sin\theta}{9.8 \text{ m/s}^2} - \dfrac{2v^2}{9.8 \text{ m/s}^2} = 0$

(e) $\quad v^2 = v_0^2 - 2(9.8 \text{ m/s}^2)\left(v_0\tan\theta - \dfrac{1}{2}(9.8 \text{ m/s}^2)t^2\right)$

(f) $\quad t = \dfrac{v^2 + (4.9 \text{ m/s}^2)x}{(3.0 \text{ m/s}^2)x}$

Solution

IDENTIFY, SET UP, AND EXECUTE: For each of the six equations, carefully examine the units of each term in the equation. Equations (*a*), (*c*), and (*d*) are dimensionally correct; however, (*b*), (*e*), and (*f*) are incorrect. The far-right term in equation (*b*) has units of (m/s), while the other two terms have units of (m). In equation (*e*), the left term inside the rightmost set of parentheses (the term $v_0 \tan\theta$) has units of (m/s), which, when combined with the (m/s^2) outside of the left parenthesis, would result in units of (m^2/s^3). The other three terms in the equation have units of (m^2/s^2). The fraction in equation (*f*) has no units, while the left-hand side has units of (s).

Thus, an error exists in each of the three equations ((*b*), (*e*), and (*f*)), since the units on the two sides of the equation do not agree. The next step would be to recheck our derivation to locate the source of the mistake.

EVALUATE: Dimensional analysis is a powerful technique to help keep you from making errors. Catching the three errors in this problem would save time while reducing confusion. Always check your units!

3: Maximum and minimum magnitudes of a vector

Given vector $\vec{A}$ with magnitude 1.3 N and vector $\vec{B}$ with magnitude 3.4 N, what are the minimum and maximum magnitudes of $\vec{A} + \vec{B}$?

Solution

IDENTIFY, SET UP, AND EXECUTE: The maximum magnitude is achieved when the two vectors are parallel and point in the same direction. The minimum magnitude is achieved when the two vectors are parallel and point in opposite directions. (The vectors are then called antiparallel.)

For parallel vectors, the magnitude is the sum of their magnitudes, 4.7 N in this case. For antiparallel vectors, the magnitude is the difference of their magnitudes, 2.1 N here.

EVALUATE: This example helps illustrate the fact that vectors do not add like ordinary scalar numbers. They do not subtract like scalar numbers either. The magnitude of $\vec{A} + \vec{B}$ for any arbitrary alignment of the two vectors must lie between 2.1 N and 4.7 N.

4: Finding parallel and perpendicular vectors

If you are given two vectors, how can you determine whether the vectors are parallel or perpendicular?

Solution

IDENTIFY, SET UP, AND EXECUTE: The cross product of two parallel vectors is zero. The dot product of two perpendicular vectors is zero. By taking the cross and dot products of the two vectors, you will determine whether they are parallel or perpendicular.

EVALUATE: There are circumstances in which you cannot easily identify parallel and perpendicular vectors, such as vectors lying in the *xyz* plane and that are given in terms of their components. Here, the best way to identify their orientation is to take dot and cross products.

Problems

1: Convert knots to m/s

A yacht is traveling at 18.0 knots. (One knot is 1 nautical mile per hour.) Find the speed of the yacht in m/s.

Solution

IDENTIFY AND SET UP: We'll use a series of conversion factors to solve this problem. Appendix E has 1 nautical mile = 6080 ft and 1 mi = 5280 ft = 1.609 km. We know that 1 km = 1000 m and that 1 hour = 60 min = 60 × (60 s) = 3600 s.

EXECUTE: We apply the conversion factors to the speed in knots to solve:

$$18.0 \text{ knots} = \left(\frac{18.0 \text{ nautical miles}}{1 \text{ h}}\right)\left(\frac{6080 \text{ ft}}{1 \text{ nautical mile}}\right)\left(\frac{1.609 \text{ km}}{5280 \text{ ft}}\right)\left(\frac{1000 \text{ m}}{1 \text{ km}}\right)\left(\frac{1 \text{ h}}{3600 \text{ s}}\right) = 9.26 \text{ m/s.}$$

EVALUATE: Using a combination of several conversion factors, we have found that 18.0 knots is equal to 9.26 m/s. The last four quantities in parentheses are each equal to unity; hence, multiplying 18.0 knots by several factors of unity doesn't change the magnitude of the quantity. Crossing out the units helps prevent mistakes.

2: Finding components of vectors

Find the x and y components of the vector $\vec{A}$ in Figure 1.6. The magnitude of vector $\vec{A}$ is 26.2 cm.

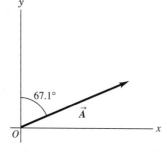

Figure 1.6 Problem 2.

Solution

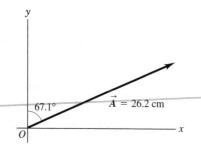

Figure 1.7 Problem 2 with components.

IDENTIFY AND SET UP: We will find the components of a vector by examining the triangle formed by the vector and the coordinate axes. Figure 1.7 shows Figure 1.6 redrawn to include the component vectors.

EXECUTE: The x component of $\vec{A}$ is located opposite the 67.1° angle; hence, we'll use the sine function:

$$A_x = A\sin 67.1° = (26.2 \text{ cm})\sin 67.1° = 24.1 \text{ cm}.$$

The y component of $\vec{A}$ is located adjacent to the 67.1° angle; thus, we'll use the cosine function:

$$A_y = A\cos 67.1° = (26.2 \text{ cm})\cos 67.1° = 10.2 \text{ cm}.$$

The vector has an x component of 24.1 cm and a y component of 10.2 cm.

EVALUATE: Finding the components of the vector required applying the sine and cosine functions. Often, but not always, the horizontal components will use cosine and the vertical components will use sine. This example illustrates an exception to that general assertion. It is important to examine a problem carefully in order to identify the proper trigonometric function for each component.

3: Vector addition

Find the vector sum $\vec{A} + \vec{B}$ of the two vectors in Figure 1.8. Express the results in terms of components.

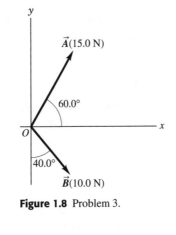

Figure 1.8 Problem 3.

Solution

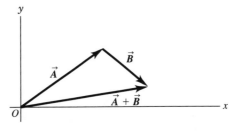

Figure 1.9 Sketch of Problem 3.

IDENTIFY AND SET UP: Figure 1.9 shows a sketch of the two vectors added together, head to tail. The sketch indicates that we should expect a resultant in the first quadrant, with positive x and y components. We will add the vectors by adding their x and y components, using the Cartesian coordinate system provided.

EXECUTE: We find the components of the vectors by examining the triangles made by the vectors and their components. For $\vec{A}$,

$$A_x = A\cos 60.0° = (15.0\,\text{N})\cos 60.0° = 7.50\,\text{N},$$
$$A_y = A\sin 60.0° = (15.0\,\text{N})\sin 60.0° = 13.0\,\text{N}.$$

For $\vec{B}$,

$$B_x = B\sin 40.0° = (10.0\,\text{N})\sin 40.0° = 6.43\,\text{N},$$
$$B_y = -B\cos 40.0° = -(10.0\,\text{N})\cos 40.0° = -7.66\,\text{N}.$$

We can now sum the components:

$$R_x = A_x + B_x = 7.50\,\text{N} + 6.43\,\text{N} = 13.9\,\text{N},$$
$$R_y = A_y + B_y = 13.0\,\text{N} - 7.66\,\text{N} = 5.34\,\text{N}.$$

The resultant vector has an x component of 13.9 N and a y component of 5.34 N.

EVALUATE: The resultant vector has positive components and resides in the first quadrant, as expected. Note how the components of the two vectors include both sine and cosine terms (i.e., vector A's x component includes the cosine component, and vector B's x component includes the sine component). This results from how the vectors' angles were given: vector A's angle was with respect to the horizontal axis and vector B's angle was with respect to the vertical axis. It is critical not to automatically associate all horizontal components with the cosine and all vertical components with the sine.

Practice Problem: Find the magnitude and direction of the resultant vector. *Answer:* The magnitude is 14.9 N, and its direction is 21.0° above the positive x-axis.

4: Determine displacement on a lake

Marie paddles her canoe around a lake. She first paddles 0.75 km to the east, then paddles 0.50 km 30° north of east, and finally paddles 1.0 km 50° north of west. Find the resulting displacement from her origin.

Solution

IDENTIFY: Displacement is a vector indicating change in position. The displacement vector points from the starting point of a journey to the endpoint. If we represent each of the three segments of the journey as a vector, the displacement vector is the sum of the three vectors. The goal is to find the sum of the three displacement vectors.

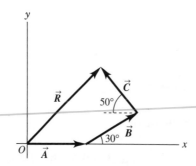

Figure 1.10 Problem 4.

SET UP: Figure 1.10 shows a sketch of the three displacement segments (labeled $\vec{A}$, $\vec{B}$, and $\vec{C}$) and the resultant displacement vector ($\vec{R}$). We will add the three vectors, using the Cartesian coordinate system in the figure.

EXECUTE: We find the components of the vectors by examining the triangles made by the vectors and their components. For $\vec{A}$, there is only a horizontal component:

$$A_x = A = 0.75 \text{ km},$$
$$A_y = 0.$$

For $\vec{B}$,

$$B_x = B\cos 30° = (0.50 \text{ km})\cos 30° = 0.443 \text{ km},$$
$$B_y = B\sin 30° = (0.50 \text{ km})\sin 30° = 0.250 \text{ km}.$$

For $\vec{C}$,

$$C_x = -C\cos 50° = -(1.0 \text{ km})\cos 50° = -0.643 \text{ km},$$
$$C_y = C\sin 50° = (1.0 \text{ km})\sin 50° = 0.766 \text{ km}.$$

The x component is negative, as it points to the west. We can now sum the components:

$$R_x = A_x + B_x + C_x = 0.75 \text{ km} + 0.443 \text{ km} - 0.643 \text{ km} = 0.550 \text{ km},$$
$$R_y = A_y + B_y + C_y = 0 \text{ km} + 0.250 \text{ km} + 0.766 \text{ km} = 1.016 \text{ km}.$$

The resultant displacement vector has an x component of 0.55 km and a y component of 1.02 km. We can express the displacement vector in terms of magnitude and direction. To find the magnitude, we use the Pythagorean theorem:

$$R = \sqrt{R_x^2 + R_y^2} = \sqrt{(0.550 \text{ km})^2 + (1.016 \text{ km})^2} = 1.16 \text{ km}.$$

The inverse tangent gives us the angle:

$$\theta = \tan^{-1}\frac{R_y}{R_x} = \tan^{-1}\frac{1.016 \text{ km}}{0.550 \text{ km}} = 61.6°.$$

The resultant displacement vector has a magnitude of 1.16 km and points 61.6° above the positive x-axis.

EVALUATE: Marie paddled a total of 2.25 km, only to end up 1.16 km away from her starting point. This shows how the magnitude of a vector sum can be smaller than the sum of the magnitudes of the individual vectors. Note that we carried an extra significant figure through the calculations and rounded off only in the final step.

5: Finding the dot product

Find the dot product of vectors $\vec{A}$ and $\vec{B}$ if $\vec{A} = 5.0\hat{i} + 2.3\hat{j} - 6.4\hat{k}$ and $\vec{B} = 12.0\hat{i} - 4.7\hat{j} + 9.3\hat{k}$.

Solution

IDENTIFY AND SET UP: We are given the x, y, and z components of the two vectors. We will find the scalar product by using the component form of the dot-product relation.

EXECUTE: The dot product is the sum of the products of the components of the vectors:

$$\vec{A} \cdot \vec{B} = A_x B_x + A_y B_y + A_z B_z.$$

For our two vectors,

$$\vec{A} \cdot \vec{B} = (5.0)(12.0) + (2.3)(-4.7) + (-6.4)(9.3)$$
$$= 60 - 10.8 - 59.5$$
$$= -10.3.$$

The dot product is -10.3.

EVALUATE: The result is a negative number, indicating that the projection of one vector onto the other points in the direction opposite that of the other. The vectors are not perpendicular, since the dot product is not zero.

We could have attempted to sketch these vectors, but since they are in three-dimensional xyz space, it is difficult to represent them accurately on a two-dimensional page. Building intuition in two-dimensional space helps us when we work in three-dimensional space.

6: Finding the cross product

Find the cross product $\vec{A} \times \vec{B}$, given $\vec{A} = 5.0\hat{i} + 2.3\hat{j} - 6.4\hat{k}$ and $\vec{B} = 12.0\hat{i} - 4.7\hat{j} + 9.3\hat{k}$.

Solution

IDENTIFY AND SET UP: We are given the x, y, and z components of the two vectors. We will find the cross product by using the component form of the cross-product relation.

EXECUTE: The three components of the cross product are given by various products of the components of the two vectors:

$$C_x = A_y B_z - A_z B_y$$
$$C_y = A_z B_x - A_x B_z$$
$$C_z = A_x B_y - A_y B_x.$$

For this problem, the components are

$$C_x = (2.3)(9.3) - (-6.4)(-4.7) = (21.4) - (30.1) = -8.7$$
$$C_y = (-6.4)(12.0) - (5.0)(9.3) = (-76.8) - (46.5) = -123.3$$
$$C_z = (5.0)(-4.7) - (2.3)(12.0) = (-23.5) - (27.6) = -51.1.$$

The cross product is $\vec{C} = -8.7\hat{i} - 123.3\hat{j} - 51.1\hat{k}$.

EVALUATE: The result is a vector, as is expected for the cross product. The vectors are not parallel, since the cross product is not zero.

Practice Problem: Find the magnitude of the resultant vector. *Answer:* The magnitude is 133.8.

7: Review of simultaneous equations

Solve the following expressions for T_A and T_B.

$$27T_A + 13T_B = 0$$
$$32T_A + 52T_B = 22.$$

Solution

IDENTIFY AND SET UP: Both of the expressions involve two unknowns, so we cannot find a solution by using only one equation. We will multiply the first equation by 32, multiply the second by 27, and then subtract the second equation from the first.

EXECUTE: Multiplying the first expression by 32 and the second expression by 27 gives

$$864T_A + 416T_B = 0$$
$$864T_A + 1404T_B = 594$$

Subtracting the second equation from the first gives

$$864T_A + 416T_B - 864T_A - 1404T_B = 0 - 594,$$
$$988T_B = 594,$$
$$T_B = \frac{594}{988} = 0.601.$$

Substituting the value for T_B back into either expression to find T_A yields

$$32T_A + 52(0.601) = 22,$$
$$T_A = -0.289.$$

The two equations together result in $T_A = -0.29$ and $T_B = 0.60$.

EVALUATE: An alternative solution would be to write T_B in terms of T_A, using the first expression, and then substitute for T_B into the second expression. This gives the same result. You may choose the method you prefer and may end up applying both to particular classes of problems.

If we encounter *three* unknowns in an expression, how many equations will we need to solve for each unknown simultaneously? Three equations will be needed to solve for the three unknown quantities.

8: Review of the quadratic formula

The position of a ball tossed in the air depends on the initial speed of the ball and the time elapsed and is given by

$$y = v_0 t - (4.9 \text{ m/s}^2)t^2,$$

where v_0 is the initial speed and t is the time elapsed. For a ball tossed with an initial speed of 30.0 m/s, find the time(s) when the ball is at a height of 12.5 m.

Solution

IDENTIFY AND SET UP: We recognize that the equation is quadratic, since it has a t^2 term, a t term, and a constant term. If we try to rewrite the equation in terms of t alone, we find that we cannot easily isolate the t term. We will employ the quadratic formula to solve the problem.

EXECUTE: We rewrite the equation, substituting the given values:

$$12.5 \text{ m} = (30.0 \text{ m/s})t - (4.9 \text{ m/s}^2)t^2.$$

The quadratic formula requires that the equation be written as $ax^2 + bx + c = 0$, so we rearrange terms to yield

$$(-4.9 \text{ m/s}^2)t^2 + (30.0 \text{ m/s})t + (-12.5 \text{ m}) = 0.$$

From this rearrangement, we see that $a = -4.9 \text{ m/s}^2$, $b = 30.0 \text{ m/s}$, and $c = -12.5 \text{ m}$. The solutions of the quadratic equation are

$$x = \frac{-b \pm \sqrt{b^2 - 4ac}}{2a}.$$

Substituting our values into the quadratic equation gives

$$t = \frac{-(30.0 \text{ m/s}) \pm \sqrt{(30.0 \text{ m/s})^2 - 4(-4.9 \text{ m/s}^2)(-1.25 \text{ m})}}{2(-4.9 \text{ m/s}^2)}.$$

Multiplying out the terms and canceling the units produces

$$t = \frac{-(30.0 \text{ m/s}) \pm \sqrt{(900.0 - 245.0)(\text{m}^2/\text{s}^2)}}{-9.8 (\text{m/s}^2)} = \frac{-30.0 \pm 25.59}{-9.8} \text{ s} = 0.445 \text{ s}, 5.67 \text{ s}.$$

There are two times when the ball is at a height of 12.5 m: 0.445 s and 5.67 s. These are, respectively, when the ball is rising to its maximum height and when it is falling from its maximum height.

EVALUATE: You must learn to recognize quadratic equations. Once you identify a quadratic equation, the solution is straightforward (although it requires careful algebra). Quadratic equations result in two solutions, and you must be able to interpret their meanings. In this case, the two solutions corresponded to the upward and downward motion of the ball. You may need only one of the solutions for your situation. If neither solution seems reasonable, then you should check your work. We'll encounter quadratic equation problems again in Chapter 2.

CAUTION **Watch units!** It is important to check units every time you write equations. If we found incorrect units when we solved for time, we would have discovered a mistake that would have been quickly corrected.

Problem Summary

The problems in this chapter represent a foundation that you will use throughout your physics course. Common elements make up good problem-solving techniques, including

- Identifying a procedure to find the solution.
- Making a sketch when no figure is provided.
- Adding appropriate coordinate systems to the sketch.
- Identifying the known and unknown quantities in the problem.
- Finding appropriate equations to solve for the unknown quantities.
- Checking for consistency of units in derived equations.
- Evaluating results to check for inconsistencies.

We will see how these techniques apply to a wide variety of problems as we progress. Although they may seem cumbersome right now, they will help you solve the problems you encounter.

2 Motion along a Straight Line

Summary

We will introduce *kinematics*, the study of an object's motion, or change of position with time, in this chapter. Motion includes *displacement*, the change in position of an object; *velocity*, the rate of change of position with respect to time; and *acceleration*, the rate of change of velocity with respect to time. We introduce average velocity and average acceleration as changes over a time interval and instantaneous velocity and instantaneous acceleration as changes over an infinitely short time interval. We'll learn relationships between displacement, velocity, and acceleration and see how they are modified for freely falling objects. We'll restrict ourselves to motion along a straight line, or one-dimensional motion, in this chapter and expand our examination to motion in two or three dimensions in the next chapter. This is our first step into understanding mechanics, the study of the relationships among force, matter, and motion that we'll cover in the upcoming chapters.

Objectives

After studying this chapter, you will understand

- The definitions of kinematic variables for position, velocity, and acceleration.
- How to calculate and interpret average and instantaneous velocities.
- How to calculate and interpret average and instantaneous accelerations.
- How to apply the equations of motion for constant acceleration.
- How to apply equations of constant acceleration to freely falling objects.
- How to analyze motion when acceleration is not constant.

Concepts and Equations

Term Description	
Average Velocity	A particle's average x-velocity v_{av-x} over a time interval Δt is its displacement Δx divided by the time interval Δt: $$v_{av-x} = \frac{x_2 - x_1}{t_2 - t_1} = \frac{\Delta x}{\Delta t}.$$ The SI unit of velocity is meters per second (m/s).
Instantaneous Velocity	A particle's instantaneous velocity is the limit of the average velocity as Δt goes to zero, or the derivative of position with respect to time. The x component is defined as $$v_x = \lim_{t \to \infty} \frac{\Delta x}{\Delta t} = \frac{dx}{dt}.$$ The term *velocity* refers to the instantaneous velocity.
Average Acceleration	The average x acceleration of a particle over a time interval Δt is the change in the x component of velocity, $\Delta v_x = v_{2x} - v_{1x}$, divided by the time interval Δt: $$a_{av-x} = \frac{v_{2x} - v_{1x}}{t_2 - t_1} = \frac{\Delta v_x}{\Delta t}.$$ The SI unit of acceleration is meters per second per second (m/s^2).
Instantaneous Acceleration	A particle's instantaneous acceleration is the limit of the average acceleration as Δt goes to zero, or the derivative of the velocity with respect to time. The x component is defined as $$a_x = \lim_{t \to \infty} \frac{\Delta v_x}{\Delta t} = \frac{dv_x}{dt}.$$ The term *acceleration* refers to the instantaneous acceleration.
Motion with Constant Acceleration	When the x acceleration is constant, position, x velocity, acceleration, and time are related by $$x = x_0 + v_{0x}t + \tfrac{1}{2}a_x t^2$$ $$v_x = v_{0x} + a_x t$$ $$v_x^2 = v_{0x}^2 + 2a_x(x - x_0)$$ $$x - x_0 = \left(\frac{v_{0x} + v_x}{2}\right)t.$$
Freely Falling Body	A freely falling body is a body that moves under the influence of the gravity. The acceleration due to gravity is denoted by g, is directed downwards, and has a value of 9.8 m/s^2 near the surface of the earth.
Motion with Varying Acceleration	When the acceleration is not constant, we can find the position and velocity as a function of time by integrating the acceleration function: $$x = x_0 + \int_0^t v_x dt$$ $$v_x = v_{0x} + \int_0^t a_x dt.$$

Conceptual Questions

1: Velocity and acceleration at the top of a ball's path

A ball is tossed vertically upward. (a) Describe the velocity and acceleration of the ball just before it reaches the top of its flight. (b) Describe the velocity and acceleration of the ball at the instant it reaches the top of its flight. (c) Describe the velocity and acceleration of the ball just after it reaches the top of its flight.

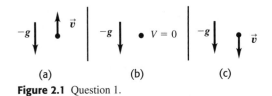

Figure 2.1 Question 1.

Solution

IDENTIFY, SET UP, AND EXECUTE: Figure 2.1 shows the three time frames we will examine. During its flight, the ball undergoes acceleration due to gravity. The initial velocity is directed upward, slowing to zero at the top of the flight. Then the velocity increases downward.

PART (A): The velocity is directed upward and is very small just before the top of the flight. The acceleration due to gravity is directed downward.

PART (B): The velocity is zero at the top of the flight. The acceleration due to gravity remains constant and is directed downward. The acceleration has caused the velocity to decrease from its small positive value in part (a) to zero.

PART (C): The velocity is directed downward and is very small just after the top of the flight. The acceleration due to gravity remains constant and directed downward. The acceleration has caused the velocity to increase downward from zero in part (b).

EVALUATE: The acceleration due to gravity causes a change in the velocity of the ball during its flight. The ball starts with an upward velocity, which slows, drops to zero, and then increases downward. The acceleration due to gravity is constant throughout the motion. The velocity is zero for an instant at the top, changing from slightly upward to slightly downward around this instant.

CAUTION   **Gravity doesn't turn off!** A common misconception is that there is no acceleration due to gravity at the top of an object's trajectory, but how would gravity know to turn off at that instant?

2: Comparing two cyclists

The position-versus-time graph depicting the paths of two cyclists is shown in Figure 2.2. (a) Do the cyclists start from the same position? (b) Are there any times that they have the same velocity? (c) What is happening at the intersection of lines *A* and *B*?

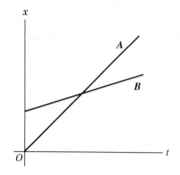

Figure 2.2 Question 2.

Solution

IDENTIFY, SET UP, AND EXECUTE PART (A): We find the starting location by examining the position when time is zero (i.e., by looking at the *x*-intercept). At *t* = 0, the two cyclists are at different locations.

PART (B): The velocity is found by examining the slope of the position-versus-time graph. The slopes of the two lines are different; hence, the cyclists never have the same velocity.

PART (C): At the intersection of lines *A* and *B*, both cyclists are at the same position at the same time. At this point, cyclist *A* is passing cyclist *B*, since cyclist *A* started closer to the origin and has a higher velocity.

EVALUATE: These three questions show only a small part of what can be learned from graphs, which offer a parallel representation of physical phenomena. The interpretation of graphs is an important tool in physics and, indeed, science in general.

3: Interpreting a position-versus-time graph

Figure 2.3 shows a position-versus-time graph of the motion of a car. Describe the velocity and acceleration during segments *OA*, *AB*, and *BC*.

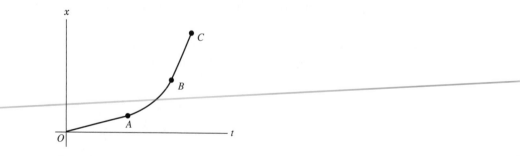

Figure 2.3 Question 3.

Solution

IDENTIFY, SET UP, AND EXECUTE: Velocity is the change in position with respect to time and acceleration is the change in velocity with respect to time. We can describe the velocity by examining the slope of the position-versus-time graph, and we can describe the acceleration by noting how the velocity changes.

In segment *OA*, the slope is positive and constant, indicating that the velocity is positive and constant. With constant velocity, there is no acceleration.

In segment *AB*, the slope is increasing smoothly, indicating that the velocity is increasing. There must be acceleration in order for the velocity to increase.

In segment *BC*, the slope is again constant, indicating that the velocity is constant. This velocity is greater in magnitude than the velocity in segment *OA,* since the slope is larger. With constant velocity, there is no acceleration.

EVALUATE: This question illustrates how we can describe the velocity and acceleration from the position-versus-time graph.

4: A falling ball

A ball falls from the top of a building. If the ball takes time t_A to fall halfway from the top of the building to the ground, is the time it takes to fall the remaining distance to the ground equal to, greater than, or smaller than t_A?

Solution

IDENTIFY, SET UP, AND EXECUTE: We can break the problem up into two segments: the first half and the second half. In the first segment, the falling ball starts with an initial velocity of zero. In the second segment, the ball has acquired velocity, so it has an initial velocity. The time to complete the second segment must be shorter than t_A.

EVALUATE: If you watch a ball fall, you should be able to see that it spends more time in the first half of the motion than in the second half. We can also look at the equation for the position of a falling body:

$$y = y_0 + v_{0y}t + \tfrac{1}{2}a_y t^2.$$

For the first half of the motion, the velocity term is zero; for the second half, it is not zero. Given equal time and equal acceleration, a segment with an initial velocity will cover a larger distance, or cover the same distance in a shorter time.

Problems

1: Throwing a ball upward

Robert throws a ball vertically upward from the edge of a 150-m-tall building. The ball falls to the ground 9.5 s after leaving Robert's hand. Assume that the ball leaves Robert's hand when it is 2.0 m above the roof of the building. Find the initial velocity of the ball and the time the ball reaches its maximum height.

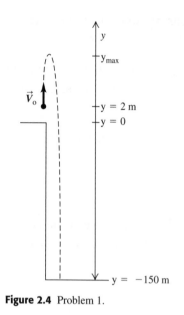

Figure 2.4 Problem 1.

Solution

IDENTIFY: The ball undergoes constant acceleration due to gravity, so we will use the constant-acceleration kinematics equation to solve the problem.

SET UP: Figure 2.4 shows a sketch of the problem. Once thrown, the ball has an initial upward velocity and will undergo downward gravitational acceleration.

We ignore effects due to the air. A vertical coordinate system is shown in the diagram, with the origin located at the edge of the building and positive values directed upward.

EXECUTE: We first determine the initial velocity of the ball. We know the initial and final positions, times, and accelerations of the ball; therefore, we use the equation for position as a function of time:

$$y = y_0 + v_{0y}t + \tfrac{1}{2}a_y t^2.$$

The initial position of the ball (y_0) is $+2.0$ m, the final position (y) is -150 m (the ground is below the edge of the building), the acceleration is $-g$, and the time is 9.5 s. Solving for the initial velocity v_{0y} gives

$$v_{0y} = \frac{y - y_0 - \tfrac{1}{2}a_y t^2}{t}.$$

Substituting the given values yields

$$v_{0y} = \frac{(-150\text{ m}) - (2.0\text{ m}) - \frac{1}{2}(-9.8\text{ m/s}^2)(6.5\text{ s})^2}{(6.5\text{ s})} = 8.5\text{ m/s}.$$

The initial velocity of the ball is 8.5 m/s. The value is positive, indicating that the initial velocity is directed upward. To find the time taken to reach the maximum height, we know that the velocity at that height is momentarily zero, so we can use the equation for velocity as a function of time:

$$v_y = v_{0y} + a_y t.$$

We now solve for the time t when the velocity v_y is zero

$$t = \frac{v_y - v_{0y}}{a_y} = \frac{(0 - 8.5\text{ m/s})}{(-9.8\text{ m/s}^2)} = 0.87\text{ s}.$$

The ball reaches its maximum height 0.87 s after leaving Robert's hand.

EVALUATE: This is a straightforward application of constant-acceleration kinematics. We identified the known and unknown quantities and substituted into appropriate equations to find the unknown quantities.

PRACTICE PROBLEM: Find the maximum height y_{max} of the ball. *Answer:* $y_{max} = 5.7$ m above the top of the building.

2: Dropping a stone from a moving helicopter

A helicopter is ascending at a constant rate of 18 m/s. A stone falls from the helicopter 12 s after it leaves the ground. How long does it take for the stone to reach the ground?

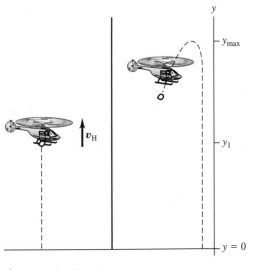

Figure 2.5 Problem 2.

Solution

IDENTIFY: There are two segments of the stone's motion: (1) moving upward with the helicopter at constant velocity and (2) free fall after the stone breaks free of the helicopter. To solve the problem, we will apply the constant-acceleration kinematics equations to the two segments, using the final quantities from the first segment as the initial quantities in the second.

SET UP: The two segments of the stone's motion are sketched in Figure 2.5. As it moves upward with the helicopter, the stone has a constant velocity v_{y0}. When it breaks loose and begins to fall freely, the stone has an initial velocity that is the same as the helicopter's and undergoes acceleration due to gravity. We need to know the position, velocity, and time at the end of the first segment to solve for the second segment. The velocity and time are given in the statement of the problem.

We ignore effects due to the air. A vertical coordinate system is shown in the diagram, with the origin located on the ground and positive values directed upward.

EXECUTE: The position is found from the equation for position as a function of time with zero acceleration:

$$y = y_0 + v_{0y}t.$$

The initial position is zero (the helicopter starts at the ground) and the helicopter is ascending, so v_{y0} is $+18$ m/s and the time is 12 s. Substituting yields

$$y = y_0 + v_{0y}t = 0 + (18 \text{ m/s})(12 \text{ s}) = 216 \text{ m}.$$

For the second segment, the initial position is 216 m, the initial velocity is $+18$ m/s, the final position is zero, and the acceleration is $-g$. The equation for position as a function of time with constant acceleration can be used to find the time:

$$y = y_0 + v_{0y}t + \tfrac{1}{2}a_y t^2.$$

Substituting values gives

$$0 = (216 \text{ m}) + (18 \text{ m/s})t + \tfrac{1}{2}(-9.8 \text{ m/s}^2)t^2.$$

We cannot solve this equation directly for t, so we resort to the quadratic equation. In this case, $a = -4.9$ m/s^2, $b = 18$ m/s, and $c = 216$ m. The result is given by

$$t = \frac{-b \pm \sqrt{b^2 - 4ac}}{2a}.$$

Substituting and solving yields

$$t = \frac{-(18 \text{ m/s}) \pm \sqrt{(18 \text{ m/s})^2 - 4(-4.9 \text{ m/s}^2)(216 \text{ m})}}{2(-4.9 \text{ m/s}^2)} = -5.1 \text{ s}, +8.7 \text{ s}.$$

The positive solution, 8.7 s, corresponds to the time the stone hits the ground. The stone hits the ground 8.7 s after falling from the helicopter, or 20.7 s after the helicopter originally left the ground.

EVALUATE: We applied the equations for motion with constant acceleration to each of the two segments in this problem, using the results from the first part as input into the second part. The negative solution of the quadratic equation corresponds to the time the stone would have left the ground, assuming that it was thrown from the ground. Because this aspect of the motion doesn't apply to our problem, we ignore that solution.

3: Avoiding a ticket

A speed trap is made by placing two pressure-sensitive tracks across a highway, 150 m apart. Suppose you are driving and you notice the speed trap and begin slowing down the instant you cross the first track. If you are moving at a rate of 42 m/s and the speed limit is 35 m/s, what must your minimum acceleration be in order for you to have an average speed within the speed limit by the time your car crosses the second track?

Solution

IDENTIFY: For the average speed over the interval to be under the speed limit, the final speed at the second track must be less than the speed limit. We will use the kinematic equations to find the acceleration necessary to avoid a speeding ticket.

SET UP: A sketch of the problem is shown in Figure 2.6. We determine the final speed by writing the average speed in terms of the initial and final speeds, setting the average speed to the speed limit, and solving for the final speed. Once we determine the final speed, we find the acceleration from the kinematics equations.

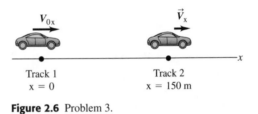

Track 1
x = 0

Track 2
x = 150 m

Figure 2.6 Problem 3.

EXECUTE: The average speed (for constant acceleration) is

$$v_{av,x} = \frac{v_{0x} + v_x}{2}.$$

Substituting and solving for the final speed gives

$$v_x = 2_{av,x} - v_{0x} = 2(35 \text{ m/s}) - (42 \text{ m/s}) = 28 \text{ m/s}.$$

The final speed must be 28 m/s in order for the average speed to be 35 m/s. We use the equation for velocity as a function of position with constant acceleration, or

$$v_x^2 = v_{0x}^2 + 2a_x(x - x_0).$$

In our coordinate system, the difference between the final and initial positions is 150 m. Substituting and solving for the acceleration gives

$$a_x = \frac{v_x^2 - v_{0x}^2}{2(x - x_0)} = \frac{(28 \text{ m/s})^2 - (42 \text{ m/s})^2}{2(150 \text{ m})} = -3.3 \text{ m/s}^2.$$

You will need to accelerate at a rate of -3.3 m/s^2 to avoid a ticket, with the minus indicating that you will need to slow down.

EVALUATE: The challenge in this problem was to recognize that we needed a final velocity that would result in the correct average velocity. A common mistake is to take the desired *average* velocity as the *final* velocity. Understanding the difference can help you avoid errors (and a ticket!).

4: Graphical solution to an accelerating car

A car undergoing constant acceleration moves 250 m in 8.5 s. If the speed at the end of the 250 m segment is 33 m/s, what was the car's speed at the beginning of the segment?

Solution

IDENTIFY: We can approach this problem in two ways. First, we can use the kinematics equations to solve the problem, but doing so will require several equations. Second, we can use the velocity-versus-time plot and solve the problem graphically. We will choose the graphical method in this case.

SET UP: A sketch of the problem is shown in Figure 2.7. We realize that there is no single kinematics equation that ties these quantities together, so we construct the velocity-versus-time graph. The graph must start with initial velocity v_0 and result in final velocity v_1. The slope of the line between the two velocities must be constant, since the car exhibits constant acceleration. The time interval between the two velocities must be the given 8.5 s, so we construct the graph shown in Figure 2.8.

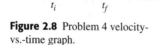

Figure 2.7 Problem 4 sketch.

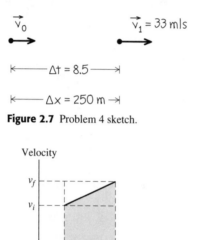

Figure 2.8 Problem 4 velocity-vs.-time graph.

Examining the graph, we realize that the area under the velocity line is the distance the car travels. We can therefore find the initial velocity by calculating the area under the curve.

EXECUTE: The area under the curve is the sum of the area of the rectangle and the area of the triangle shown in the figure. We find these areas by multiplying the time by the velocities:

$$\text{Area} = v_0 t + \tfrac{1}{2}(v_1 - v_0)t.$$

The area under the curve is just the distance the car travels, 250 m. We rewrite the equation to solve for the initial velocity v_0:

$$v_0 = \frac{2\text{Area} - v_1 t}{t} = \frac{2((250\text{ m}) - (33\text{ m/s})(8.5\text{ s}))}{(8.5\text{ s})} = 25.8\text{ m/s}.$$

The initial velocity is 25.8 m/s.

EVALUATE: The example illustrates how graphical analysis can lead to a straightforward solution. The key was to realize that the area under the curve is the distance the car traveled, or its displacement. If we were to solve the problem with kinematics equations, we would have had two unknowns (initial velocity and acceleration), requiring us to utilize two equations in the solution.

5: Two objects falling from a building

A ball is dropped from the top of a tall building. One second later, another ball is thrown from the top of the building with a velocity of 30 m/s directed vertically downwards. Will the balls ever meet? If so, when and where?

Solution

IDENTIFY: Both balls undergo acceleration due to gravity after being dropped or thrown, so we will apply constant-acceleration kinematics. We will use separate sets of kinematic equations for the two balls; ball 1 will be the dropped ball and ball 2 will be the thrown ball.

SET UP: A sketch of the problem is shown in Figure 2.9. We are interested in where the balls meet, so we will use the position equations, setting their positions equal to each other to find out whether they meet at any point in time. The coordinate system is shown in the sketch, with the origin at the top of the building and positive values directed downward.

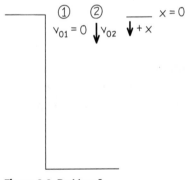

Figure 2.9 Problem 5.

EXECUTE: The equation for the position as a function of time for ball 1, the dropped ball, is

$$x_1 = x_{01} + v_{01}t + \tfrac{1}{2}a_x t^2 = \tfrac{1}{2}gt^2.$$

The equation for the position as a function of time for ball 2, the thrown ball, is

$$x_2 = x_{02} + v_{02}t' + \tfrac{1}{2}a_x t'^2 = v_{02}(t - 1\text{ s}) + \tfrac{1}{2}a_x(t - 1\text{ s})^2,$$

where we have included the initial velocity v_{02} and the replaced the time t' with $(t - 1\text{ s})$. For the two balls to meet, the two positions must be the same. We set the two equations equal to each other and solve for the time they meet:

$$\tfrac{1}{2}a_x t^2 = v_{02}(t - 1\text{ s}) + \tfrac{1}{2}a_x(t - 1\text{ s})^2$$

$$\tfrac{1}{2}gt^2 = v_{02}t - v_{02}1\text{ s} + \tfrac{1}{2}gt^2 + \tfrac{1}{2}g(-2t\text{ s}) + \tfrac{1}{2}g(1\text{ s})^2.$$

The t^2 term cancels, leaving

$$0 = v_{02}t - v_{02}1\text{ s} + \tfrac{1}{2}g(-2t\text{ s}) + \tfrac{1}{2}g(1\text{ s})^2.$$

Solving for t gives

$$t = \frac{\frac{1}{2}g(s) - v_{02}}{g(s) - v_{02}} = \frac{\frac{1}{2}(9.8 \text{ m/s}^2)(s) - (30 \text{ m/s})}{(9.8 \text{ m/s}^2)(s) - (30 \text{ m/s})} = 1.24 \text{ s.}$$

The balls meet 1.24 s after the first ball is dropped. The position of the balls at this time is

$$x_1 = \frac{1}{2}gt^2 = \frac{1}{2}(9.8 \text{ m/s}^2)(1.24 \text{ s})^2 = 7.57 \text{ m.}$$

The balls meet 7.57 m below the top of the building.

EVALUATE: We check our results and see that the balls meet 1.24 s after the first ball is dropped, or 0.24 s after the second ball is thrown. Since the balls meet after the second ball is thrown, we conclude these times represent a reasonable result. They meet 7.57 m below the top of the building. (A positive value indicates that they meet below the top of the building.) We also see that the building must be at least 7.57 m tall.

6: Nonconstant acceleration

A particle has an initial velocity of 12.0 m/s and starts at $x = 14.2$ m. It moves along the x axis and possesses an acceleration given by

$$a_x = bt^2,$$

where b is a constant equal to 3.5 m/s^4. Find the particle's velocity and position as a function of time.

Solution

IDENTIFY AND SET UP: The acceleration varies with time, so we cannot use the constant-acceleration kinematic equations. Instead we integrate the acceleration to find the velocity as a function of time and we integrate the velocity to find the position as a function of time. Both the initial position and initial velocity are zero.

EXECUTE: We begin by integrating the acceleration to find the velocity:

$$v_x = v_{0x} + \int_0^t a_x dt.$$

For our problem, v_{0x} is 12.0 m/s and a_x is given. Substituting and solving, we obtain

$$v_x = (12.0 \text{ m/s}) + \int_0^t bt^2 dt$$

$$= (12.0 \text{ m/s}) + \frac{1}{3}bt^3 \Big|_0^t$$

$$= (12.0 \text{ m/s}) + \frac{1}{3}bt^3.$$

To find the position as a function of time, we integrate the velocity:

$$x = x_0 + \int_0^t v_x dt.$$

Substituting $x_0 = 14.2$ m as given and the value of v_x that we found and solving yields

$$v_x = \int_0^t \left[(12.0 \text{ m/s}) + \tfrac{1}{3} bt^3 \right] dt$$

$$= (14.2 \text{ m}) + \left[(12.0 \text{ m/s})t + \tfrac{1}{12}bt^4 \right]\big|_0^t$$

$$= (14.2 \text{ m}) + (12.0 \text{ m/s})t + \tfrac{1}{12}bt^4.$$

We have found both the position and velocity as a function of time.

EVALUATE: We can check our integration by taking derivatives of our results. When we do, we find the original acceleration.

Try It Yourself!

Learning physics requires that you practice problems without having solutions next to the problem. To help you prepare for homework problems and exams, we have included sample problems with checkpoints to help you through them. We encourage you to try these problems on your own and refer to the checkpoints only when you get stuck. So go ahead and *Try It Yourself!*

1: Police chase

A speeder traveling at a constant speed of 100 km/hr passes a waiting police car that immediately starts from rest and accelerates at a constant 2.5 m/s². (a) How long will it take for the police car to catch the speeder? (b) How fast will the police car be traveling when it catches the speeder? (c) How far will the police car have traveled when it catches the speeder?

Solution Checkpoints

IDENTIFY AND SET UP: Constant-acceleration kinematics are appropriate in this problem. Two separate sets of kinematics equations should be used to represent the police car and the speeder. We set appropriate values equal to each other to solve the problem.

EXECUTE: The positions of the police car and speeder must be the same when the police car catches the speeder:

$$v_{0, \text{speeder}} t = \tfrac{1}{2} at^2.$$

This leads to the conclusion that the speeder is caught 22.2 s after passing the police car. Kinematics then indicates that the police car had a velocity of 200 km/hr and a position of 616 m.

EVALUATE: Would you expect the police car to accelerate at a constant rate or the speeder not to slow down? How would these changes affect the result? $t = 0$ s is also a solution of the equation. To what event does $t = 0$ s correspond?

2: Don't hit the truck

A car traveling 100 km/hr is 200 m away from a truck traveling 50 km/hr in the same direction. What minimum acceleration must the car have in order to avoid hitting the truck? Assume constant acceleration during braking.

Solution Checkpoints

IDENTIFY AND SET UP: Constant-acceleration kinematics are valid in this problem. Two sets of kinematics equations should be used to represent the car and truck separately. Choose an appropriate coordinate system.

EXECUTE: To avoid the collision with the minimum acceleration, the car and truck will meet at the same point at the same time and with the same velocity. Setting the car's and truck's position equations equal to each other gives

$$v_{0car}t + \tfrac{1}{2}at^2 = x_{0truck} + v_{0truck}t.$$

This equation has two unknowns, so you will need to set the velocities equal to each other and solve by using both relations. You should find that the magnitude of the acceleration is 0.48 m/s^2.

EVALUATE: What is the sign of the acceleration you found? Is it what you would expect for a car slowing down?

3: Throwing a ball upward

A ball is thrown vertically upward from a 125-m-high building with an initial velocity of 45 m/s. (a) What is the ball's maximum height? (b) What is its velocity as it passes the top of the building on its way down? (c) How long does it take the ball to reach the ground?

Solution Checkpoints

IDENTIFY AND SET UP: Constant-acceleration kinematics are valid in this problem. You will need equations for position as a function of time and velocity as a function of time. Choose an appropriate coordinate system.

EXECUTE (A): At the maximum height, the velocity is zero. Solving will give a height of 103 m above the building.

(b) On the way down, the velocity of the ball when it passes the top of the building will have the same magnitude as the initial velocity, but will be opposite in direction.

(c) The equation for position as a function of time can be used to find the time taken for the ball to hit the ground. The position equation will lead to a quadratic equation and a result of 11.4 s.

EVALUATE: The quadratic equation of part (c) had two roots. Why did you omit one root? What is the physical interpretation of the omitted root?

3 Motion in Two or Three Dimensions

Summary

In this chapter we expand our kinematics to motion of bodies in two or three dimensions. In doing so, we will find that we can simultaneously apply our one-dimensional kinematics equations to multiple axes independently. Displacement, velocity, and acceleration take on their vector qualities as we expand to more dimensions, requiring us to work with components of each quantity. Our new skills will allow us to investigate projectile motion and the interesting case of uniform circular motion. We will also learn to analyze motion viewed from different moving reference frames. By the end of this chapter, we will have laid a strong foundation in kinematics and will be ready to investigate the causes of motion.

Objectives

After studying this chapter, you will understand

- How to describe a body's position, velocity, and acceleration in terms of vector quantities.
- How to apply equations of motion to bodies moving in a plane.
- How to describe and analyze the motion of projectiles.
- How to analyze an object in uniform circular motion.
- How to combine components of acceleration that are parallel and perpendicular to a body's path.
- How to relate the velocities of objects to different reference frames.

Concepts and Equations

Term	Description
Position Vector	The position vector $\vec{r}$ of a point P in space is the displacement vector from the origin to P. It has components x, y, and z.
Average Velocity	The average velocity $\vec{v}_{av}$ during a time interval Δt is the displacement $\Delta \vec{r}$ divided by Δt: $$\vec{v}_{av} = \frac{\vec{r}_2 - \vec{r}_1}{t_2 - t_1} = \frac{\Delta \vec{r}}{\Delta t}.$$
Instantaneous Velocity	A body's instantaneous velocity is the derivative of $\vec{r}$ with respect to time: $$\vec{v} = \lim_{t \to \infty} \frac{\Delta \vec{r}}{\Delta t} = \frac{d\vec{r}}{dt}.$$ The instantaneous velocity has components $$v_x = \frac{dx}{dt}, \qquad v_y = \frac{dy}{dt}, \qquad v_z = \frac{dz}{dt}.$$
Average Acceleration	The average acceleration $\vec{a}_{av}$ during a time interval Δt is the change in velocity $\Delta \vec{v}$ divided by Δt: $$\vec{a}_{av} = \frac{\vec{v}_2 - \vec{v}_1}{t_2 - t_1} = \frac{\Delta \vec{v}}{\Delta t}$$
Instantaneous Acceleration	A body's instantaneous acceleration is the derivative of $\vec{v}$ with respect to time: $$\vec{a} = \lim_{t \to \infty} \frac{\Delta \vec{v}}{\Delta t} = \frac{d\vec{v}}{dt}.$$ The instantaneous velocity has components $$a_x = \frac{dv_x}{dt}, \qquad a_y = \frac{dv_y}{dt}, \qquad a_z = \frac{dv_z}{dt}.$$ The component of acceleration parallel to the velocity affects the speed of the body. The component of acceleration perpendicular to the velocity affects the body's direction of motion. A body has acceleration if either its speed or direction changes.
Projectile Motion	A body undergoes projectile motion when it is given an initial velocity and then follows a path determined entirely by the effect of a constant gravitational force. The path, or trajectory, is a parabola. The projectile's vertical motion is independent of its horizontal motion. The horizontal acceleration is zero and the vertical acceleration is $-g$. The coordinates and velocities of a projectile with an initial velocity of magnitude v_0 and direction α_0 (measured with respect to the ground) are given as a function of time by $$x = (v_0 \cos \alpha_0)t,$$ $$y = (v_0 \sin \alpha_0)t - \tfrac{1}{2}gt^2,$$ $$v_x = v_0 \cos \alpha_0,$$ $$v_y = v_0 \sin \alpha_0 - gt.$$

Uniform Circular Motion	A particle moving in a circular path of radius R and constant speed v is said to move in uniform circular motion. The particle possesses an acceleration of magnitude $$a_{\text{rad}} = \frac{v^2}{R}$$ directed toward the center of the circle. If the particle's speed is not constant in circular motion, then the radial component of acceleration remains as just given and there is also a component parallel to the path of the particle.
Relative Velocity	When a body P moves relative to a reference frame B, and B moves relative to a second reference frame A, the velocity of P relative to B is denoted by $\vec{v}_{P/B}$, the velocity of P relative to A is denoted by $\vec{v}_{P/A}$, and the velocity of B relative to A is denoted by $\vec{v}_{B/A}$. These velocities are related by $$\vec{v}_{P/A} = \vec{v}_{P/B} + \vec{v}_{B/A}.$$

Conceptual Questions

1: Velocity and acceleration at the top of a projectile's path

A projectile is launched with initial nonzero x and y velocities. (a) Describe the velocity and acceleration just before the projectile reaches the top of its trajectory. (b) Describe the velocity and acceleration at the instant the projectile reaches the top of its trajectory. (c) Describe the velocity and acceleration just after the projectile reaches the top of its trajectory.

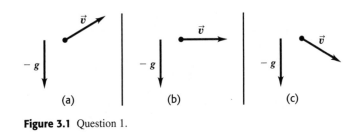

Figure 3.1 Question 1.

Solution

IDENTIFY AND SET UP: Figure 3.1 shows the three time frames we will examine. During the flight, the projectile undergoes acceleration due to gravity. The projectile's initial velocity has both x and y components; the x component remains constant while the y component is accelerated by gravity. We have to consider each component of velocity separately.

EXECUTE PART (A): The x component of velocity is constant and directed to the right, and the y component of velocity is directed upward and is very small just before the top of the flight. The acceleration due to gravity is directed downward.

PART (B): The x component of velocity is constant and directed to the right, and the y component of velocity is zero at the top of the flight. The acceleration due to gravity remains constant and directed downward.

PART (C): The x component of velocity is constant and directed to the right, and the y component of velocity is directed downward and is very small just after the top of the flight. The acceleration due to gravity remains constant and directed downward.

EVALUATE: The acceleration due to gravity causes a change in velocity during the flight, but affects only the vertical component of velocity. The ball starts with a nonzero velocity, which decreases, drops to a minimum value at the top, and then increases downward. The acceleration due to gravity is constant throughout the motion. The velocity is changing throughout the motion. Question 1 from Chapter 2 is similar to this problem.

2: Launching a marble off the edge of a table

A marble is launched off the edge of a horizontal table and lands on the floor. Draw the trajectory of the ball from the table to the floor. Draw a second line showing the trajectory of the marble if it were given a smaller initial velocity. Draw a third line showing the trajectory if the marble were given a larger initial velocity than the original initial velocity.

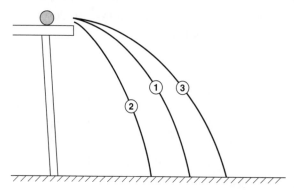

Figure 3.2 Question 2.

Solution

IDENTIFY, SET UP, AND EXECUTE: The table and marble are sketched in Figure 3.2. The initial trajectory is shown and labeled "1." The marble follows a parabolic path, starting with an initial nonzero horizontal velocity. For the smaller initial velocity, the marble also follows a parabolic path, but with a termination point closer to the edge of the table. This path is shown in the figure and is labeled "2." For the larger initial velocity, the marble again follows a parabolic path, but with a termination point farther from the edge of the table. This path is shown in the figure and is labeled "3."

EVALUATE: The paths are similar; their differences owe to the different initial velocities. How does the time the marble spends in the air compare for the three paths? All three take the same amount of time to reach the ground, as they all start with zero initial vertical velocity and fall the same distance. Since they spend the same time in the air, those with larger initial velocities reach greater horizontal distances.

3: Comparing projectiles

Figure 3.3 graphs the paths of two projectiles in the xy plane. If we ignore air resistance, how do the initial velocities compare (magnitude and direction)?

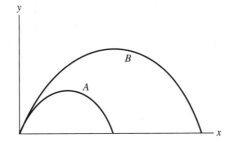

Figure 3.3 Question 3.

Solution

IDENTIFY, SET UP, AND EXECUTE: At the origin, we see that the two paths begin identically. This indicates that the initial directions of the velocities of both projectiles are the same.

The graph does not include a time axis, so we need to look to other clues to compare the magnitudes of the initial velocities. Trajectory B reaches a greater height, indicating that its initial vertical component of velocity was larger than trajectory A's initial vertical component. Therefore, trajectory B has a greater magnitude of initial velocity.

EVALUATE: Without a time axis, examine the x motion and we cannot assume that trajectory B has a greater magnitude of initial velocity.

How do the horizontal components of the initial velocities compare? Both projectiles have the same initial launch angle; therefore, the ratio of their velocity components must be the same. If the vertical component of B's velocity is larger, so must B's horizontal velocity component be larger.

4: Comparing projectiles again

Figure 3.4 shows the graph of the paths of two projectiles in the xy plane. If we ignore air resistance, how do the initial velocities compare (in magnitude and direction)? Which projectile lands first?

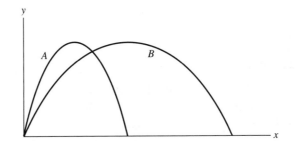

Figure 3.4 Question 4.

Solution

IDENTIFY, SET UP, AND EXECUTE: We see that the paths do not coincide at the origin: Projectile *A* has a larger launch angle.

Again, the graph does not include a time axis, so we need to look to other clues to compare the magnitudes of the initial velocities. Both trajectories reach the same maximum height, indicating that both have the same vertical velocity components. However, their initial directions were different, requiring their initial horizontal components to be different. Trajectory *B* reaches a greater horizontal distance, so must have a larger initial horizontal velocity. Therefore, trajectory *B* has a greater magnitude of initial velocity.

Since both trajectories reach the same maximum height and have the same initial vertical velocity, they must end at the same time.

EVALUATE: As an alternative analysis, we could have considered the time first and the velocity second. In that case, it might have been easier to see that the initial horizontal velocity of projectile *B* was larger because it covered more distance in the same time.

5: Falling luggage

A piece of luggage falls out of the cargo door of a airplane flying horizontally at a constant speed. In what direction should the pilot look to follow the luggage to the ground so that it can be recovered?

Solution

IDENTIFY, SET UP, AND EXECUTE: When the piece of luggage falls from the airplane, its initial velocity is the same as the plane's velocity. As it falls, the luggage accelerates in the vertical direction and its horizontal velocity remains constant (assuming no air resistance). Since the plane and the piece of luggage are moving at the same horizontal velocity, the luggage falls directly below the plane. The pilot should look straight down to see where the luggage will land.

EVALUATE: You might expect that the piece of luggage would fall behind the airplane. Now, what would cause it to fall behind the plane? For it to fall behind the plane, the luggage would have to slow down, or accelerate in a direction opposite that of its horizontal motion. Air resistance could slow down the bag, because air resistance opposes the motion of a body.

Problems

1: Water Balloon Launch

Your physics professor is walking past the physics building at a constant 3.5 m/s. You're on the third-floor balcony (25 m above the ground) of the building with your new water balloon launcher. The launcher allows you to adjust the speed of the water balloon, but you can launch the balloon only horizontally. What launch speed should be set for the balloon to land on your professor if you launch it just as she passes below? What will be your professor's horizontal distance from the building when the balloon hits her, as measured from a point on the ground directly below you?

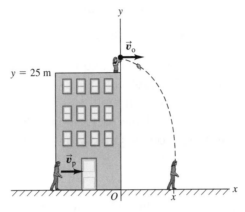

Figure 3.5 Problem 1.

Solution

IDENTIFY: Once launched, the water balloon will undergo gravitational acceleration in the vertical direction and continue with constant velocity in the horizontal direction. We will apply the constant-acceleration kinematics equations separately to the horizontal and vertical components to solve the problem.

SET UP: Figure 3.5 shows a sketch of the situation. We ignore effects due to the air. Your professor is roughly 1.7 m tall, but we'll ignore her height and determine the position where the balloon hits the ground. An xy coordinate system is shown in the diagram.

We first determine the launch speed. Since there is no acceleration in the horizontal direction, the water balloon must be launched at the same speed as your professor is walking, 3.5 m/s.

EXECUTE: To find where the balloon hits her, we find the time from the start of the vertical motion and use that to find the horizontal distance the water balloon travels as it falls. The vertical position for constant acceleration is given by

$$y = y_0 + v_{0y}t + \tfrac{1}{2}at^2.$$

We've set the origin at the ground; therefore, the initial position becomes 25 m and the final position becomes 0. The launcher imparts only a horizontal velocity, so the initial vertical velocity is zero. The acceleration is $-g$, since the positive vertical axis is directed upward. Combining these parameters gives

$$0 = 25 \text{ m} - \tfrac{1}{2}gt^2.$$

Solving for t produces

$$t = \sqrt{\frac{2(25 \text{ m})}{(9.8 \text{ m/s}^2)}} = 2.26 \text{ s}.$$

It takes 2.26 s for the balloon to fall to the ground. During this time, it is traveling with constant horizontal velocity. We find the horizontal distance it travels from the formula

$$x = x_0 + v_{0x}t.$$

Your origin is directly below your position on the balcony $(x_0 = 0)$. Substituting the horizontal velocity and time we calculated, we find the horizontal distance:

$$x = v_{0x}t = (3.5 \text{ m/s})(2.26 \text{ s}) = 7.9 \text{ m.}$$

The water balloon will hit your professor a horizontal distance 7.9 m away from your location.

EVALUATE: This is a straightforward application of two-dimensional kinematics. We solved for one component of the motion and substituted the result into the equation for the other component to arrive at the solution. Note that we solved for the vertical motion and substituted the result into the equation for the horizontal motion, the opposite order of the previous problem. Practicing solving a variety of problems will build proficiency in solving problems involving motion in a plane.

2: Hitting a Baseball in Fenway Park

You win a chance to try hitting a baseball over the "Green Monster" in Fenway Park. The Green Monster is a 37.2-ft (11.3-m)-high wall in left field of the ballpark. The left end is closest to home plate, 310 ft (94.5m) away. If you give the ball an initial speed of 33 m/s at an initial angle of 47°, by how much does the baseball clear (or miss) the top of the wall?

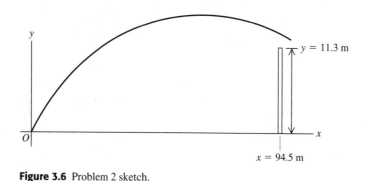

Figure 3.6 Problem 2 sketch.

Solution

IDENTIFY: The baseball has a nonzero initial velocity, undergoes acceleration due to gravity in the vertical direction, and has no acceleration in the horizontal direction. Constant-acceleration kinematics equations will be applied separately to the horizontal and vertical components to find the solution.

SET UP: We sketch the problem in Figure 3.6. We ignore effects due to the air. The ball is hit roughly 1 m or so above the ground, but we'll neglect this small distance and set the origin at ground level. An xy coordinate system that coincides with this choice is shown in the diagram.

We will solve for the time the baseball arrives at the wall by using the horizontal-position equation. Then we will substitute into the vertical-position equation to find the vertical position of the ball at the wall.

EXECUTE: The horizontal position is given by

$$x = x_0 + v_{0x}t.$$

In this case, we start at the origin $(x_0 = 0)$ and the x component of velocity includes a cosine term:

$$x = v_{0x}t = v_0 \cos\theta t.$$

We wish to find the time t when the baseball is located at the wall $(x = 94.5 \text{ m})$:

$$t = \frac{x}{v_0 \cos\theta} = \frac{(94.5 \text{ m})}{(33 \text{ m/s}) \cos 47°} = 4.20 \text{ s}.$$

After 4.20 s, the baseball's horizontal position is 94.5 m. We now find the vertical position at that time. The vertical position is given by

$$y = y_0 + v_{0y}t + \tfrac{1}{2}at^2$$

Again, we start at the origin $(y_0 = 0)$, the acceleration is directed downward (negative) and is of magnitude g, and the y-component of velocity includes a sine term:

$$y = v_0 \sin\theta t + \tfrac{1}{2}(-g)t^2.$$

We can now substitute our values into the equation to find the height of the ball:

$$y = (33 \text{ m/s})(\sin 47°)(4.20 \text{ s}) + \tfrac{1}{2}(-9.8 \text{ m/s}^2)(4.20 \text{ s})^2 = 14.9 \text{ m}.$$

At the wall, the ball's height is 14.9 m, or 3.6 m above the 11.3-m-high wall. The ball clears the Green Monster by 3.6 m!

EVALUATE: This is another straightforward application of two-dimensional kinematics. We solved for one component of the motion and substituted the result into the equation for the other component to arrive at the solution. We will follow this procedure often to solve problems involving motion in a plane.

Practice Problem: Find the x and y components of the baseball's velocity at the wall. *Answer*: $v_x = 22.5 \text{ m/s}, v_y = -17.0 \text{ m/s}$.

CAUTION   **Don't Mix x and y!** It is easy to mix up x and y components for position, velocity, and acceleration. You must label each of these carefully to ensure that you don't make mistakes. Only time is common to both the x and y components.

3: Acceleration of a Propeller Tip

The Wright Brothers' plane had a 2.4-m-long propeller that operated at a constant 350 rpm. Find the acceleration of a particle at the tip of the propeller.

Solution

IDENTIFY: This is a uniform circular motion problem; the acceleration is determined by the centripetal-acceleration formula.

SET UP: We will need to find the velocity and radius from the information provided. A diagram of the problem is shown in Figure 3.7.

To find the centripetal acceleration, we need the radius and speed of a particle on the tip of the propeller. We are given the diameter of the propeller, and dividing that in half gives the radius.

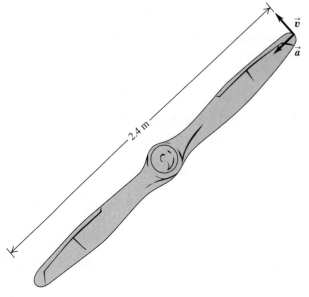

Figure 3.7 Problem 3.

EXECUTE: The speed of a particle at the end of the propeller is found by dividing the circumference at the tip of the propeller $(2\pi r)$ by the time it takes the propeller to make one revolution: (T)

$$v = \frac{2\pi r}{T}.$$

The propeller makes 350 revolutions per minute, so we find the time it takes to make 1 revolution by dividing 1 minute by 350 revolutions:

$$T = \frac{1 \text{ min}}{350 \text{ rev}} = \frac{60 \text{ s}}{350 \text{ rev}} = 0.171 \text{ s/rev.}$$

The propeller takes 0.171 s to make one revolution. We can now find the velocity:

$$v = \frac{2\pi r}{T} = \frac{2\pi (1.2 \text{ m})}{0.171 \text{ s}} = 44.1 \text{ m/s.}$$

The centripetal acceleration is then

$$a_{rad} = \frac{v^2}{r} = \frac{(44.1 \text{ m/s})^2}{(1.2 \text{ m})} = 1620 \text{ m/s}^2.$$

The centripetal acceleration of a particle on the tip of the propeller is 1620 m/s². This is equivalent to 165 times the acceleration due to gravity!

EVALUATE: We have found the magnitude of the acceleration in this problem. Acceleration is a vector, so where does it point? The acceleration is directed toward the center of the propeller, perpendicular to the velocity.

We did not include gravity in this problem. The particle is affected by attraction toward the ground throughout its motion; however, the resulting acceleration is very small compared with the centripetal acceleration.

4: Paddling across a River

You wish to paddle north across a 350-m-wide river. The river has a 1.2 m/s current from east to west, and you can paddle at a steady 1.5 m/s pace. In what direction should you paddle, and how long will it take you to cross the river?

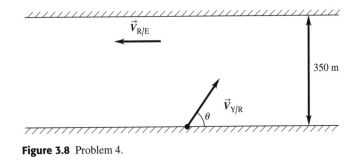

Figure 3.8 Problem 4.

Solution

IDENTIFY: You will need to paddle into the river current to compensate for the river's moving your boat downstream as you cross.

SET UP: Figure 3.8 shows a sketch of the situation. We use relative velocities to solve the problem. The direction in which you must paddle is determined by setting north to be the direction of your resulting relative velocity with respect to the earth.

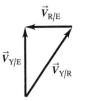

Figure 3.8 Problem 4.

EXECUTE: Figure 3.9 combines your velocity with respect to the river $(v_{Y/R})$ with the river current's velocity with respect to the earth $(v_{R/E})$ to form your relative velocity with respect to the earth $(v_{Y/E})$:

$$\vec{v}_{Y/E} = \vec{v}_{Y/R} + \vec{v}_{R/E}.$$

For you to land directly across from your starting point, the direction of $v_{Y/E}$ must be northward. Therefore, the x component of $v_{Y/R}$ must be equal and opposite to $v_{R/E}$. We find the direction in which you should paddle by equating those two magnitudes:

$$(\vec{v}_{Y/R})_x = v_{R/E},$$

$$(\vec{v}_{Y/R})_x = v_{Y/R}\sin\theta = v_{R/E}.$$

$$\theta = \sin^{-1}\left(\frac{v_{R/E}}{v_{Y/R}}\right) = \sin^{-1}\left(\frac{1.2 \text{ m/s}}{1.5 \text{ m/s}}\right) = 53°.$$

You will need to paddle 53° east of north to follow a northward path. The time it will take is found from the y component of the displacement. You are traveling at constant velocity, so the vertical component of the displacement is

$$y - y_0 = \left(v_{Y/E}\right)_y t.$$

$\left(v_{Y/E}\right)_y$ is the magnitude of $v_{Y/E}$, because $v_{Y/E}$ has only a y component. $\left(v_{Y/E}\right)_y$ must also be equal to the y component of $v_{Y/R}$. Solving for time gives

$$t = \frac{y - y_0}{\left(v_{Y/E}\right)_y} = \frac{y - y_0}{v_{Y/E}} = \frac{y - y_0}{\left(v_{Y/R}\right)_y} = \frac{y - y_0}{v_{Y/E}\cos\theta} = \frac{350 \text{ m}}{\left(1.5 \text{ m/s}\right)\cos\left(53°\right)} = 390 \text{ s}.$$

It will take you 390 s to paddle across the river.

EVALUATE: When you paddle across a river perpendicular to its flow, your relative velocity with respect to the earth is always less than your velocity with respect to the river. It also takes longer to cross a river that has a current compared with a calm river when your path is perpendicular to the river. The next practice problem lets you compare the time required to cross a calm river with the time you just found in dealing with a river that has a current.

Practice Problem: How long would it take to paddle across the same river with no current? *Answer*: 230 s.

Try It Yourself!

1: Ball thrown from a cliff

A boy throws a ball from a cliff at an angle of 30.0° above the horizontal with an initial velocity of 10.0 m/s. The ball lands 100.0 m from the base of the cliff. (a) How high is the cliff? (b) What is the time of flight of the ball? (c) What is the velocity of the ball just before impact?

Solution Checkpoints

IDENTIFY AND SET UP: Constant-acceleration kinematics equations should be applied separately to the horizontal and vertical components of the ball's motion. The initial velocity should be broken down into x and y components. There is acceleration only in the vertical direction.

EXECUTE: Equations for the x and y components of position and velocity can be found. The x position equation can be solved for time and substituted into the y position equation to find the height of the cliff. Doing this leads to

$$y = x\tan\theta - \tfrac{1}{2}g\left[\frac{x}{v_0\cos\theta}\right]^2,$$

from which you will find that the cliff is 595 m high. You can then substitute that number into the x position equation to find that the time is 11.5 s.

To find the velocity just before impact, you can find the velocity components from the kinematic equations. Knowing the components, you can find the magnitude and direction of the velocity. The velocity is 108 m/s, directed 85° below the x-axis.

EVALUATE: We see that the velocity just before impact is almost straight down. Can it ever be exactly straight down?

2: Kicking a soccer ball

A soccer ball is kicked 25 m in the horizontal direction. What is its initial velocity if it reaches a maximum height of 6.0 m?

Solution Checkpoints

IDENTIFY AND SET UP: Constant-acceleration kinematics equations should be applied separately to the horizontal and vertical components of the ball's motion. Draw the ball's path and choose an appropriate coordinate system.

EXECUTE: The initial y component of velocity can be found from the formula

$$v_v^2 = v_{0y}^2 + 2a_y\Delta y.$$

This gives an initial vertical velocity of 10.8 m/s. To find the initial horizontal component of velocity, we determine the flight time from the vertical motion and combine that with the horizontal distance traveled. The result is an initial velocity of 15.6 m/s at an angle of 44° above the horizontal.

EVALUATE: By this point, you have seen many kinematics problems. Can you summarize your problem-solving techniques?

3: Archer and arrow

An archer shoots an arrow into the air at an angle of 30° above the horizontal. It lands on a building 100.0 m away at a height of 20.0 m. What was the initial speed of the arrow?

Solution Checkpoints

IDENTIFY AND SET UP: Can constant-acceleration kinematics be used in this case? What assumptions do you need to make?

EXECUTE: The x and y position equations can be rearranged to yield

$$y = x\tan\theta - \tfrac{1}{2}g\left[\frac{x}{v_0\cos\theta}\right]^2.$$

Rewriting the equation and the x and y positions at the building gives $v_0 = 41$ m/s.

EVALUATE: How can you check your result? How do you know that it is reasonable?

4 Newton's Laws of Motion

Summary

We will define dynamics—the study of the relationship of motion to forces—in this chapter. Newton's laws of motion will lay the foundation for our studies and link forces to acceleration, which we investigated in the previous chapters. We will define force, mass, and weight and apply them to problems. We will use our knowledge of vectors to better understand forces and construct free-body diagrams. By the end of this chapter, we will have built a problem-solving framework that we will apply in the next chapter.

Objectives

After studying this chapter, you will understand

- The concept of force and why it is a vector quantity.
- How to identify forces acting on a body.
- How to find the resultant force acting on an object by summing multiple forces.
- Newton's three laws of motion.
- The relation between net force, mass, and acceleration.
- How to recognize an inertial frame of reference, in which Newton's laws are valid.
- How to use a free-body diagram to represent forces acting on an object.
- How to use the free-body diagram as a guide in writing force equations for Newton's laws.

Concepts and Equations

Term	Description
Force	A force is a quantitative measure of the interaction between two objects, represented by a vector. The SI unit of force is the newton (N). One newton equals 1 kilogram-meter per second squared.
Combining Forces	The vector sum of forces acting on a body is the resultant, denoted $\vec{R}$: $$\vec{R} = \vec{F}_1 + \vec{F}_2 + \vec{F}_3 + \cdots = \sum \vec{F}.$$ The effect of many forces acting on a body can be captured by the resultant force. This principle is called **superposition of forces**.
Contact Force	A contact force is a force between two objects touching at a surface. A contact force has two components: a component perpendicular to the surface (the normal force) and a component parallel to the surface (the frictional force).
Normal Force	The normal force is the component of a contact force between two objects that is perpendicular to their common surface. The normal force is denoted by $\vec{n}$.
Friction Force	The friction force is the component of a contact force between two objects that is parallel to their common surface. The friction force is denoted by $\vec{f}$. Friction forces often act to resist the sliding of an object.
Tension Force	A tension force is conveyed by the pull of a rope or cord and is denoted by $\vec{T}$.
Newton's First Law	Newton's first law states that when the vector sum of forces acting on an object is zero, the object is in equilibrium and has zero acceleration. The object will remain at rest or move with constant velocity when no net force acts upon it. The law is valid only in inertial reference frames.
Inertial Reference Frame	An inertial reference frame is a reference frame in which Newton's laws are valid. A common example of a *non*inertial reference frame is that of an accelerating airplane.
Newton's Second Law	Newton's second law of motion states that an object which is not in equilibrium is acted upon by a net force and accelerates. The acceleration is given in vector form by $$\sum \vec{F} = m\vec{a},$$ where m is the object's mass, which characterizes the inertia of the object. Newton's second law can also be written in component form as $$\sum F_x = ma_x \qquad \sum F_y = ma_y \qquad \sum F_z = ma_z.$$
Weight	An object's weight is the gravitational force exerted on the object by the earth or another astronomical body and is denoted by $\vec{w}$. The magnitude of an object's weight is equal to the product of its mass and the magnitude of the acceleration due to gravity: $$w = mg.$$

Newton's Third Law	Newton's third law states that, for two interacting bodies A and B, each exerts a force on the other of equal magnitude and opposite in direction, or $$\vec{F}_{A \text{ on } B} = -\vec{F}_{B \text{ on } A}.$$
Free-Body Diagram	A diagram showing all forces acting *on* an object. The object is represented by a point; the forces are indicated by vectors. A free-body diagram is useful in solving problems involving forces.

Conceptual Questions

1: Winning a Tug-of-War

In a tug-of-war shown in Figure 4.1, how does the force applied to the rope by the losing team compare with the force applied to the rope by the winning team? Is the magnitude of the force applied by the losing team less than, greater than, or equal to the magnitude of the force applied by the winning team? How does the winning team win?

Figure 4.1 Question 1.

Solution

IDENTIFY, SET UP, AND EXECUTE The free-body diagram in Figure 4.2 will guide our analysis. Each team experiences four forces: the tension due to the tug-of-war rope (T), a frictional force with the ground (f), the weight of the team (w), and the normal force exerted upward from the ground (N). The subscripts indicate the winning (w) and losing (l) teams. Examining the diagrams, we see that the tension forces must be an action–reaction pair—hence the explicit notation. Therefore, by Newton's third law, the tensions must be equal. The force applied by the losing team on the rope must be of the same magnitude as, but opposite in direction to, the force applied by the winning team on the rope.

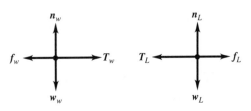

Figure 4.2 Question 1 free-body diagrams.

The vertical forces will not influence the horizontal interaction, so we look at the remaining force to determine how the winning team wins. The frictional forces must not be equal: The winning team exerts a larger frictional force on the ground than the losing team does in order to accelerate the losing team across the centerline.

EVALUATE We see that free-body diagrams and Newton's third law were crucial in our solution. The free-body diagram helped reduce the complexity of the problem and helped show that the frictional force was responsible for the win. After establishing that the tensions were an action–reaction pair, we saw from Newton's third law that the tensions were equal and opposite.

Note that we assumed that the frictional force was between the team and the ground and that each team was able to grip the rope without sliding. There is also a frictional force between the teams' hands and the rope. Differences between the hand and rope frictional forces could also have led to the win.

2: Flying groceries

What force causes a bag of groceries to fly forward when you come to an abrupt stop in a car?

Solution

IDENTIFY, SET UP, AND EXECUTE Suppose that, before you come to an abrupt stop, you are moving at a constant velocity. Then no net force must act on you, the car, or the bag of groceries, according to Newton's first law. As you cause an abrupt stop by hitting the brakes, you increase the frictional force between the car and the road, creating a net force on the car. When you brake, the force on the bag of groceries doesn't change, so the bag of groceries continues at its initial velocity. (We're assuming that the frictional force between the bag of groceries and the car seat is small.) Therefore, *no force* causes the bag of groceries to fly forward when you come to an abrupt stop in a car.

EVALUATE The solution may seem a bit illogical, for consider how the situation would appear to someone outside of the car: The bag of groceries continues moving at a constant velocity after the brakes are applied. This scenario should be more plausible and is a clearer way to imagine the situation.

This is one example of a noninertial frame of reference. The slowing car has negative acceleration and hence is an accelerated frame of reference. Newton's laws don't apply to noninertial frames of reference, so we cannot apply our new force techniques to this problem.

From inside the car, you may try to explain the situation by invoking a "force of inertia." This would be a fictitious force, however, and must be avoided. All of the forces we've encountered (and all of those we will later encounter) arise from known interactions.

3: Does an Apple Accelerate When Placed on a Table?

An apple is placed on a table. Can we describe the apple as having an acceleration of 9.8 m/s² toward earth and a second acceleration of 9.8 m/s² upward due to the table, thus resulting in a net acceleration of zero?

Solution

IDENTIFY, SET UP, AND EXECUTE We have seen how forces can cause accelerations, have heard $F = ma$ often, and know that an object's weight is mg, so it may appear logical to replace forces with mass times acceleration in equations. However, Newton's laws apply to combining forces, not accelerations. Newton's second law states that a net force on an object will lead to an acceleration equal to the net force divided by the object's mass.

EVALUATE This question points up a common misconception about accelerations and forces. At times, replacing forces with mass times acceleration may lead to the same results as following the correct procedures, but doing so often leads to confusion. An object that is stationary is not accelerating, because there is no *net* force acting on the object.

4: Forces and Moving Objects

Does a force cause an object to move? Does a moving object "have" a force?

Solution

IDENTIFY, SET UP, AND EXECUTE A force does not necessarily cause an object to move. Your textbook is acted upon by gravity when placed on a desk, but it does not move. A *net* force can cause an object to acquire velocity through acceleration.

An object moving at a constant velocity has no net force acting on it; therefore, the fact that an object is in motion does not indicate that a force is acting upon it. The fact that an object is *accelerating,* however, would certainly indicate that at least one force is acting upon it.

EVALUATE Acceleration and motion are *not* equivalent. Acceleration is motion during which the velocity changes over time. An object can also have a *constant* velocity, which is motion without acceleration. You must carefully distinguish between motion and acceleration in order to grasp physics.

5: Definition of equilibrium

Can a moving object be in equilibrium?

Solution

IDENTIFY, SET UP, AND EXECUTE Equilibrium occurs when the net force acting on an object is zero. Newton's first law states that objects with no net force acting on them remain at rest or continue with constant velocity. An object moving at constant velocity is in equilibrium.

EVALUATE *Equilibrium* has a precise definition in physics, even though the word may have connotations of a stationary object. Physics relies upon precise definitions to build representations of physical processes. You must apply physics definitions carefully to build your physics understanding.

Problems

1: Combining several forces to find the resultant

A mover uses a cable to drag a crate across the floor as shown in Figure 4.3. The mover provides a 300 N force and pulls the cable at an angle of 30.0°. The crate weighs 500 N, and the floor provides a 350 N normal force on the crate and opposes his pull with a 150 N frictional force. Find the resulting force acting on the crate. Will the crate accelerate?

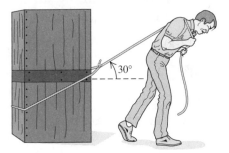

Figure 4.3 Problem 1.

Solution

IDENTIFY We will combine the forces acting on the crate to find the net force. If the net force is not zero, then there will be acceleration.

SET UP We find the resultant force by adding the forces acting on the crate. Four forces act on the crate: the tension force due to the mover's pull (T), the crate's weight (w), and the normal force (n) and friction force (f) due to the floor. We represent the four forces as vectors in the free-body diagram of the crate in Figure 4.4.

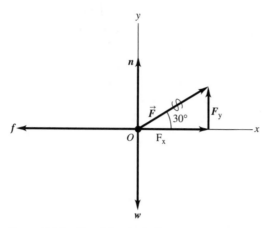

FIGURE **4.4** Problem 1 free-body diagram.

We have added an xy coordinate system to the free-body diagram as the forces act in two dimensions. We've also resolved the tension force into its x and y components.

EVALUATE We add the four forces together by adding their components, writing separate equations for the x and y components. There are two x components, due to the horizontal components of the tension force and the friction force:

$$\sum F_x = T\cos 30° + (-f).$$

The x component of the tension force is to the right and is thus assigned a positive value, while the friction force is to the left and is assigned a negative value. We proceed to the y components. There are three y components, one due to the normal force, a second due to the weight of the crate, and, finally, the vertical component of the tension force:

$$\sum F_y = n + (-w) + T\sin 30°.$$

The y component of the tension force and the normal force are directed upward and are assigned positive values, while the weight of the crate is directed downward and is assigned a negative value. We now substitute the values for the variables to find the net force along both axes:

$$\sum F_x = T\cos 30° + (-f) = (300\,\text{N})\cos 30° + (-150\,\text{N}) = +110\,\text{N},$$
$$\sum F_y = n + (-w) + T\sin 30° = (350\,\text{N}) + (-500\,\text{N}) + (300\,\text{N})\sin 30° = 0\,\text{N}.$$

The resultant force on the crate has an x component of $+110$ N and no y component, (i.e., the resulting force is horizontal and points to the right). There is also a resulting acceleration of the crate to the right, as there is a net force.

EVALUATE This is a typical force problem in which we have used our vector addition skills to find the resultant force. We see that there is no net force in the vertical direction; therefore, the crate remains on the bed of the truck.

PRACTICE PROBLEM At what rate does the crate accelerate? *Answer*: 2.2 m/s^2.

2: Using Newton's second law to find the mass of a cruise ship

A tugboat pulls a cruise ship out of port. (See Figure 4.5.) You estimate the acceleration by noting that the tugboat takes 60 s to move the cruise ship 100 m, starting from rest. If the tugboat exerts 3×10^6 N of thrust, what is the mass of the cruise ship? Ignore drag due to the water, and assume that the tugboat accelerates uniformly.

FIGURE **4.5** Problem 2.

Solution

IDENTIFY We will use Newton's second law to find the mass of the cruise ship, given the net force and acceleration acting on the ship.

SET UP The problem tells us the net force provided by the tugboat, and the acceleration can be determined from the kinematics information. We ignore drag or friction with the water, so the only horizontal force acting on the cruise ship is due to the tugboat.

EXECUTE Newton's second law relates the net force to the mass and resulting acceleration:

$$\sum F_x = ma.$$

The net force acting on the cruise ship is 3×10^6 N. The acceleration is found from the equation for position as a function of time with constant acceleration:

$$x = x_0 + v_0 t + \tfrac{1}{2}at^2.$$

Here, the initial velocity is zero and we take the initial position to be zero. Substituting these values into the equation gives

$$x = \tfrac{1}{2}at^2.$$

Solving for the acceleration yields

$$a = \frac{2x}{t^2}.$$

Replacing the distance and time with the given values produces

$$a = \frac{2x}{t^2} = \frac{2(100 \text{ m})}{(60 \text{ s})^2} = 0.056 \text{ m/s}^2.$$

We now use Newton's second law to find the mass. Solving for the mass gives

$$m = \frac{F}{a} = \frac{(3 \times 10^6 \text{ N})}{(0.056 \text{ m/s}^2)} = 54{,}000{,}000 \text{ kg} = 54 \text{ kilotonnes.}$$

Our estimate shows that the cruise ship has a mass of 54 kilotonnes ($1 \text{ kilotonne} = 10^6 \text{ kg}$). More correctly, the cruise ship has a mass of 50 kilotonnes, as the values stated in the problem have only one significant figure.

EVALUATE This problem shows how we can combine Newton's law with observations to make interesting conclusions about the mass of an object.

3: Drawing free-body diagrams

Draw a free-body diagram for each of the following situations:

(a) A box slides down a smooth ramp. (See Figure 4.6.)

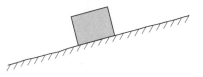

FIGURE **4.6** Problem 3a

(b) A box slides down a rough ramp. (See Figure 4.7.)

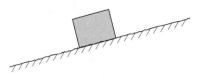

FIGURE **4.7** Problem 3b

(c) A block is placed on top of a crate, and the crate is placed on a horizontal surface. (See Figure 4.8.) Draw a free-body diagram of the crate.

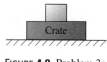

FIGURE **4.8** Problem 3c

(d) A block is placed on top of a crate, and the crate is pulled horizontally across a rough surface. (See Figure 4.9.) The surface between the crate and the block is rough, and the block is held at rest by a string. Draw a free-body diagram of the crate.

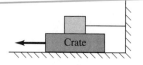

FIGURE **4.9** Problem 3d

Solution

IDENTIFY We will draw free-body diagrams that show all of the forces acting on the object, representing the forces as vectors.

SET UP The first step is to identify the object and then find the forces acting on it. We'll look at the contact tension, normal and frictional forces, and the noncontact gravitational force.

EXECUTE In part (a), there is no friction, since the ramp is smooth. The only contact force acting on the box is the normal force due to the ramp. Gravity also acts on the box. The free-body diagram includes two forces acting on the box: the normal (n) force, directed perpendicular to the ramp; and the weight (w) of the box, directed downward. The free-body diagram of the box is shown in Figure 4.10.

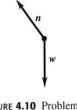

FIGURE **4.10** Problem 3a
free-body diagram

In part (b), there is friction, since the ramp is rough. The contact forces acting on the box are the normal and frictional forces due to the ramp. Gravity also acts on the box. The free-body diagram includes three forces acting on the box: the normal force (n), directed perpendicular to the ramp; the frictional force (f), directed upward along the ramp (opposing the motion of the box); and the weight (w) of the box, directed downward. The free-body diagram of the box is shown in Figure 4.11.

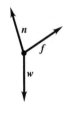

FIGURE **4.11** Problem 3b
free-body diagram

In part (c), two contact forces act on the crate: the normal force due to the surface and the normal force due to the block. There are no frictional forces, as neither the crate nor the block is moving. Gravity acts on the box. The free-body diagram includes three forces acting on the crate: the normal force due to the surface ($n_{surface}$), directed upward; the normal force due to the block (n_{block}), directed downward; and the weight (w), of the crate, directed downward. The free body diagram of the crate is shown in Figure 4.12.

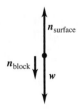

FIGURE **4.12** Problem 3c
free-body diagram

In part (d), five contact forces act on the crate: the normal forces due to the surface and the block, the frictional forces due to the surface and the block, and the tension force provided by the pull. The sixth force acting on the crate is gravity. The free-body diagram includes six forces acting on the crate: the normal force due to the surface $(n_{surface})$, directed upward; the normal force due to the block (n_{block}), directed downward; the frictional forces due to the surface $(f_{surface})$ and the block (f_{block}), both directed to the right; the tension force (T), directed to the left; and the weight (w), of the crate, directed downward. The free-body diagram for the crate is shown in Figure 4.13.

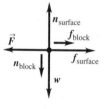

FIGURE **4.13** Problem 3d
free-body diagram

EVALUATE Drawing free-body diagrams should become second nature to you. We will see their importance when we solve problems in Chapter 5. Free-body diagrams help catch mistakes by identifying all the forces acting on an object, as well as help identify action–reaction pairs.

4: Tension in a string connecting two blocks

Two blocks are connected by a massless string, as shown in Figure 4.14. A cable is attached to the upper block and is pulled upward with a 250 N force. Find the tension in the string connecting the two blocks. The upper block has a mass of 7.5 kg and the lower block has a mass of 12 kg.

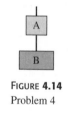

FIGURE **4.14**
Problem 4

Solution

IDENTIFY We will use Newton's laws to solve this problem.

SET UP We cannot determine whether the system is in equilibrium or accelerating from the statement of the problem; therefore, we do not know whether to apply Newton's first law for a system in equilibrium or Newton's second law for an accelerating system. Our first step, therefore, will be to determine whether the system is in equilibrium or accelerating. Then we will apply the appropriate one of Newton's laws to find the tension in the string.

We'll use free-body diagrams to solve the problem. We can determine whether the blocks are accelerating by considering the two blocks as one system. The left panel of Figure 4.15 shows a free-body diagram of the system with the two blocks combined. We can find the tension in the string by considering the two blocks separately. The middle and right panels of Figure 4.15 show the free-body diagrams of the two blocks separately. The top block is designated A, the bottom block B, to reduce confusion.

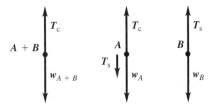

FIGURE **4.15** Problem 4 free-body diagram

The forces in the free-body diagrams are identified by their magnitudes. The combined diagram includes the tension of the cable (T_{cable}) and the weight of the two blocks (w_{A+B}). The other diagrams also include the tension of the string (T_{string}), and the weights of the blocks $(w_A$ and $w_B)$. The upward-pointing vectors will be taken to be positive, the downward-pointing vectors negative.

EXECUTE To determine whether the blocks are accelerating, we examine the net force acting on them. From the left-hand free-body diagram, we see that there are two forces, the tension of the cable and the weight of the blocks, acting on the combination of blocks:

$$\sum F_y = T_{cable} + (-w_{A+B}).$$

The weight is the combined mass of the blocks times the gravitational constant. The net force is found by replacing the weight and tension by the given values:

$$\sum F_y = T_{cable} + (-g(m_A + m_B)) = 250 \text{ N} + (-(9.8 \text{ m/s}^2)(7.5 \text{ kg} + 12 \text{ kg})) = 58.9 \text{ N}.$$

The net force is not zero; therefore, the blocks are accelerating. We find the acceleration from Newton's second law applied to the combined blocks:

$$\sum F_y = (m_A + m_B)a.$$

Solving for the acceleration yields

$$a = \frac{\sum F_y}{(m_A + m_B)} = \frac{58.9 \text{ N}}{(7.5 \text{ kg} + 12 \text{ kg})} = 3.02 \text{ m/s}^2.$$

We now apply Newton's second law to the lower block to find the tension in the string. Two forces are acting on the lower block: the tension due to the string (upward) and gravity (downward). Hence,

$$\sum F_y = T_{string} + (-m_B g) = m_B a.$$

Solving for the tension in the string and substituting the value for acceleration gives

$$T_{string} = m_B g + m_B a = m_B(g + a) = (12 \text{ kg})(9.8 \text{ m/s}^2 + 3.02 \text{ m/s}^2) = 150 \text{ N}.$$

The tension in the string is 150 N.

EVALUATE We see that the tension force due to the string is less than the tension force due to the cable. This is expected, as the string provides force to accelerate the lower block, whereas the cable provides force to accelerate both blocks. It is important not to assume that tensions are equal in problems; you must consider each tension independently.

Try It Yourself!

1: Rock suspended by wire

A rock of mass 4.0 kg is suspended by a wire. When a horizontal force of 29.4 kg is applied to the rock, it moves to the side such that the wire makes an angle θ with the vertical. Find the angle θ and the tension in the wire.

Solution Checkpoints

IDENTIFY AND SET UP The net force on the rock is zero. Three forces act on the rock. By drawing a free-body diagram, we can see how to set the horizontal and vertical components of force to zero to solve the component force equations.

EXECUTE The net horizontal and vertical forces acting on the rock are

$$\sum F_x = F_H - T\sin\theta = 0$$
$$\sum F_y = T\cos\theta - mg = 0.$$

Dividing one equation by the other leads to an angle of 37°. Substituting the angle into either equation leads to a tension of 49 N.

EVALUATE We will often break the net force into horizontal and vertical components and solve each separately, much as we did in our two-dimensional kinematics problems.

2: Tension in an elevator cable

A 1000.0 kg elevator rises with an upward acceleration equal to g. What is the tension in the supporting cable?

Solution Checkpoints

IDENTIFY AND SET UP There is a net force on the elevator, so Newton's second law will be used to find the tension.

EXECUTE The net vertical force acting on the elevator is

$$\sum F_y = T - mg = ma = mg.$$

Solving for the tension gives 19,600 N.

EVALUATE We see that the tension in the cable is larger than the force of gravity on the elevator. Does this make physical sense?

3: Acceleration in an elevator

A 100.0 kg man stands on a bathroom scale in an elevator. What is the acceleration of the elevator when the scale reading is (a) 150 kg, (b) 100 kg, and (c) 50 kg?

Solution Checkpoints

IDENTIFY AND SET UP Two forces act on the man: the force of the scale and the force of gravity. Draw a free-body diagram to guide you.

EXECUTE The net vertical force acting on the man is

$$\sum F_y = F_{scale} - mg = ma.$$

Solving for the acceleration gives (a) $a = 4.9 \text{ m/s}^2$, (b) $a = 0$, and (c) $a = -4.9 \text{ m/s}^2$.

EVALUATE In which case(s) does the man feel heavier than normal? In which case(s) does he feel lighter than normal? In which case(s) does he feel as if he has normal weight? What is the significance of the signs in answers (a) and (c)?

5 | Applying Newton's Laws

Summary

In this chapter, we will apply Newton's laws of motion to bodies that are in *equilibrium* (at rest or in uniform motion) and to bodies that are *not in equilibrium* (in accelerated motion). We'll develop a consistent problem-solving strategy that utilizes a free-body diagram to identify the relevant forces acting on a body. We'll also expand our catalog of forces by quantifying contact forces and friction forces, as well as examine forces on bodies in uniform circular motion. By the end of the chapter, you will have built a foundation for solving equilibrium and nonequilibrium problems involving any combination of forces, including forces that we'll discover in later chapters.

Objectives

After studying this chapter, you will understand

- How to efficiently represent forces acting on a body by using a free-body diagram.
- How to use the free-body diagram as a guide in writing force equations for Newton's laws.
- How to use Newton's first law to solve problems involving bodies in equilibrium.
- How to use Newton's second law to solve problems involving accelerating bodies.
- How to apply contact forces and various friction forces to a variety of situations.
- How to recognize action–reaction pairs and use Newton's third law to quantify their magnitudes.
- How to apply Newton's laws of motion to bodies moving in uniform circular motion.
- How to use Newton's laws to solve problems proficiently.

Concepts and Equations

Term	Description
Using Newtons' First Law	A body in equilibrium (either at rest or moving with constant velocity) is acted upon by no net force: The vector sum of the forces acting on the object must be zero according to Newton's first law of motion, $\sum \vec{F} = 0$. When solving equilibrium problems, one starts with free-body diagrams, finds the net forces along two perpendicular components, and then solves by using the equations $$\sum F_x = 0 \quad \sum F_y = 0.$$
Applying Newton's Second Law	A body that is acted upon by a nonzero net force accelerates. The acceleration is given by Newton's second law of motion, $$\sum \vec{F} = m\vec{a}.$$ When solving nonequilibrium problems, one identifies the forces acting on the body with free-body diagrams, finds the net forces along two perpendicular components, and determines the equations of motion given by $$\sum F_x = ma_x \quad \sum F_y = ma_y$$ In problems with multiple interacting bodies, it may be necessary to apply Newton's laws to each body individually and solve the equations simultaneously.
Friction Force	A friction force is that component of the contact force between two objects which is parallel to the surfaces in contact. Friction forces, denoted by $\vec{f}$, are generally proportional to the normal force and include kinetic friction forces (when the objects move relative to each other), static friction forces (when there is no motion between the objects), viscosity and drag forces (for motion involving liquids and gases), and rolling frictional forces (for rolling objects).
Kinetic Friction Force	The kinetic friction force is the friction force between two objects moving relative to each other and is generally proportional to the normal force between the objects. The proportionality constant is the coefficient of kinetic friction (μ_k), which depends on the objects' surface characteristics and has no units. The direction of the kinetic friction force is always opposite the direction of motion. Mathematically, $$f_k = \mu_k n.$$
Static Friction Force	The static friction force is the friction force between two objects that are not moving relative to each another. The maximum static friction is generally proportional to the normal force between the objects, where the proportionality constant is the coefficient of static friction (μ_s). Often, μ_s is greater than μ_k for a given surface. The static friction can vary from zero to the maximum value; its magnitude depends on the component of the applied forces parallel to the surface. The direction of the static frictional force is opposite that of the parallel component of the net applied force. Mathematically, $$f_s \leq \mu_s n.$$
Forces in Circular Motion	For a body in uniform circular motion, the acceleration is directed toward the center of the circle. The motion is determined by Newton's second law, $$\sum \vec{F} = m\vec{a}.$$ The body's acceleration is $$a_{rad} = \frac{v^2}{R} = \frac{4\pi^2 R}{T^2}.$$

Conceptual Questions

1: Finding errors in a free-body diagram

Two weights are suspended from the ceiling and each other by ropes as shown in Figure 5.1a. A free-body diagram is shown in Figure 5.1b for the upper block (*A*). Find the error in the free-body diagram and draw the correct diagram.

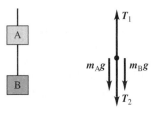

Figure 5.1 Question 1.

Solution

IDENTIFY, SET UP, AND EXECUTE Three forces act on block *A:* two tension forces due to the ropes and the gravitational force on block *A*. The gravitational force on block *B* has been incorrectly included in the diagram. Block *B* is not in direct contact with block *A;* only the rope is in contact with block *A*. The corrected free-body diagram is shown in Figure 5.2.

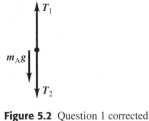

Figure 5.2 Question 1 corrected free-body diagram.

EVALUATE When drawing free-body diagrams, one must include only those forces acting *on* the object. Identifying the forces acting on an object is necessary to apply Newton's laws correctly.

2: Investigation of the normal force

For which of the following figures is the normal force not equal to the object's weight?

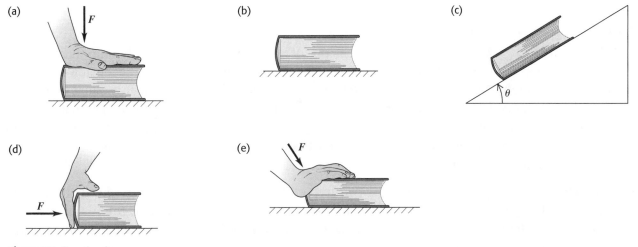

Figure 5.3 Question 2

Solution

IDENTIFY, SET UP, AND EXECUTE The normal force is not equal to the object's weight in figures (a), (c), and (e). In (a), the normal force is equal to the book's weight plus the force pushing down on the book. In (c), the book's weight is directed downward and the normal force is directed upward and to the left, perpendicular to the ramp's surface. Here, the normal force is equal to the weight multiplied by the cosine of the ramp angle. In (d), the normal force is equal to the book's weight, but the applied force is along the surface; thus, it does not affect the vertical forces. In (e), a component of the applied force is parallel to the normal force, thus increasing the normal force by the amount of that component. The two forces are directed downward.

EVALUATE Often, the normal force is not equal to an object's weight. A common mistake initially encountered in force problems is assuming that the normal force is always equal to some object's weight. You must analyze all problems carefully to determine the proper normal force.

3: Acceleration and tension in blocks connected by a rope

Consider the situation shown in Figure 5.3. Cart A is placed on a table and is connected to block B by a rope that passes over a frictionless pulley.

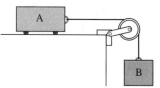

Figure 5.4 Question 3.

(a) How does the acceleration of cart A compare with that of block B?

Solution

IDENTIFY, SET UP, AND EXECUTE Both objects must accelerate at the same rate, since they are connected by the rope (as long as the rope doesn't stretch). To get a better intuitive grasp of this statement, note that cart A will move 10 cm when block B moves 10 cm. If block B moves the 10 cm in 1 second, then cart A moves 10 cm in 1 second; their velocities are the same. If block B's velocity changes by 2 m/s in 1 second, then block A's velocity must change by 2 m/s in 1 second; their accelerations are the same. We say that the rope constrains both objects to accelerate at the same rate.

(b) How does the tension force acting on cart A compare with the weight of block B as the system accelerates?

Solution

IDENTIFY, SET UP, AND EXECUTE The tension force in the string is constant along the string, so the tension force is the same on block B as it is on cart A. Therefore, we compare the tension at block B with block B's weight. Newton's second law tells us that the net force on an object is equal to its mass times its acceleration. Two forces act on block B: B's weight and the tension force. The net force on block B is its weight minus the tension force. This net force must be equivalent to the acceleration multiplied by block B's mass; therefore, the tension must be less than the weight. The tension force acting on cart A is less than the weight of block B.

EVALUATE Solving this problem gives us two important results that we'll apply repeatedly to later problems: First, objects connected by a rope are constrained to have the same magnitude of acceleration; second, the tension force in a rope connected to an object is not always equal to the object's weight if the object is accelerating.

4: What can a hanging ball indicate

A ball hangs on a string attached to the top of a box, as shown in Figure 5.4. The box is placed on a horizontal truck bed and the truck moves over a flat roadway. You observe the ball and find that it remains in the position shown in the figure for a long time. By looking only inside the box, what can be determined about the truck's motion?

Figure 5.5 Question 4.

Solution

IDENTIFY, SET UP, AND EXECUTE We see that the ball has swung to the left, so we may suspect that the truck is moving. Let's look at the forces acting on the ball: to investigate the motion. Two forces act on the ball: gravity and the tension force due to the string. Figure 5.5 shows the free-body diagram.

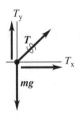

Figure 5.6 Question 4
free-body diagram.

We see that the tension force has components in both the vertical and horizontal direction. The vertical component of the tension force must be equivalent to the force of gravity, since the box moves horizontally and there is no vertical acceleration. There is only one horizontal force—the horizontal component of the tension force—so there is horizontal acceleration to the right.

We can conclude that the truck is accelerating to the right in the horizontal direction and not at all in the vertical direction. We cannot determine the velocity of the truck; the horizontal velocity components could be zero or nonzero. For example, the truck could be moving to the left with decreasing velocity or it could be accelerating to the right from rest. Both of these motions would result in the ball's being swung to the left.

EVALUATE This problem shows that the absence of a net force results in zero acceleration of an object and the presence of a net force results in a nonzero acceleration of an object. We cannot determine the velocity of an object by knowing only its acceleration; we need additional information.

Practice Problem: How would the ball appear if the box moved with constant velocity? *Answer:* The ball would hang vertically, since no net force would act on it.

5: Motion of a box on a rough surface

A constant horizontal force is applied to a box on a rough floor. With a 15 N applied force, the box begins to slide. What is the motion of the box after it begins to slide, assuming that the applied force remains constant?

Solution

IDENTIFY, SET UP, AND EXECUTE Before the box slides, there is static friction. Once it begins to slide, the static friction becomes kinetic friction. Kinetic friction is smaller in magnitude than static friction; therefore, the applied force must be larger than the kinetic friction force, and the box accelerates.

EVALUATE This problem helps illustrate the fact that kinetic friction is generally less than static friction. The reason is that the coefficient of kinetic friction is less than the coefficient of static friction.

6: Frictional forces

A box is placed on a rough floor. When you push horizontally against the box with a 20 N force, the box just begins to slide. What is the magnitude of the frictional force when you push against the box with a 10 N force? With a 15 N force?

Solution

IDENTIFY, SET UP, AND EXECUTE Since the box just begins to slide when the 20 N force is applied, the maximum static friction force is 20 N. When you push against the box with a 10 N force, you are pushing with less than the maximum static friction force. By Newton's third law, the box must push back with the same force; therefore, the static friction force must have a magnitude of 10 N. For the same reason, when you push with a 15 N force, the static friction force has a magnitude of 15 N.

EVALUATE Static friction varies from zero to its maximum value. The static friction force equals the net force acting against the friction force. Be careful not to assume that static friction is always at its maximum.

7: A vertical frictional force

A block is placed against the vertical front of an accelerating cart as shown in Figure 5.6. What condition must hold in order to keep the block from falling?

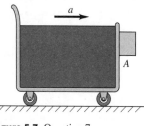

Figure 5.7 Question 7.

Figure 5.8 Question 7 free-body diagram

Solution

IDENTIFY, SET UP, AND EXECUTE The free-body diagram shown in Figure 5.7 indicates that three forces act on the box: gravity (mg, downwards), the normal force due to the cart (n, to the right), and the friction force at the block–cart surface (fs). To keep the block from falling, the friction force must be equal and opposite to the gravitational force. The condition for equilibrium in the vertical direction gives

$$\sum F_y = f_s - mg = 0, \qquad f_s = mg.$$

The friction force must be due to static friction, in order to prevent the block from moving. The static friction force is given by

$$f_s \leq \mu_s n.$$

The gravitational force can then be related to the normal force and the coefficient of static friction:

$$mg \leq \mu_s n.$$

The block accelerates to the right with acceleration a, so we can use Newton's second law to find an expression for the normal force:

$$\sum F_x = n = ma.$$

Replacing the normal force with mass and acceleration gives

$$mg \leq \mu_s ma, \qquad g \leq \mu_s a.$$

$$\mu_s \geq \frac{g}{a}.$$

The last inequality tells us that the coefficient of static friction must be equal to or greater than the gravitational constant divided by the cart's acceleration.

EVALUATE Problems that initially appear complicated often have relatively straightforward solutions. Following a consistent problem-solving procedure helps identify the key points that you will need to solve the problem.

8: Turning while riding in a car

As you make a right turn in your car, what pushes you against the car door?

Solution

IDENTIFY, SET UP, AND EXECUTE As your car turns, your body tends to continue moving in a straight line; therefore, you push up against the car door. So you can say that *no force* actually pushes you against the door and your body tries to maintain a constant velocity due to Newton's first law. Once the car door comes in contact with you, it pushes you in the direction of the turn, accelerating you to the right.

EVALUATE From within the car, you may wonder what force pushes you to the side. However, since the car is turning to the right, it is accelerating, and the car is not an inertial reference frame. Therefore, we cannot apply Newton's laws inside the car. If we consider how the situation would appear to someone outside of the car (in an inertial reference frame), we can apply Newton's laws: You would appear to continue moving in a straight line while the car moves to the right.

9: Free-body diagram for a car on a hill

Draw a free-body diagram for a car going over the top of a round hill at a constant speed. Is there a nonzero net force acting on the car if it is moving at constant speed?

Solution

SET UP AND SOLVE Two forces act on the car at the top of the hill: the normal force due to the road and gravity. The normal force is directed upward and gravity is directed downward. Figure 5.8 shows the free-body diagram.

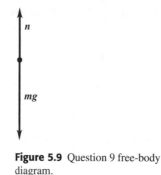

Figure 5.9 Question 9 free-body diagram.

The force vectors in the free-body diagram are not drawn to have equal length: The gravitational force is larger than the normal force. This is because there is a nonzero net force acting on the car: The car's velocity is changing direction, so the car has a centripetal acceleration. The centripetal acceleration is downward, so the net force must be downward.

There is a net force acting on the car even though the car is moving at constant speed.

REFLECT Constant speed does not necessarily imply constant velocity. You must carefully interpret problems involving circular motion and constant speed.

Problems

1: Equilibrium in two dimensions

A 322 kg block hangs from two cables as shown in Figure 5.9. Find the tension in cables A and B.

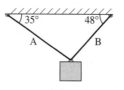

Figure 5.10 Problem 1.

Solution

IDENTIFY The block is in equilibrium, so we can apply Newton's first law to it. The cables have tensions in two dimensions, so we will have to apply the first law to two axes. The target variables are the two tension forces (labeled T_A for cable A and T_B for cable B).

SET UP The free-body diagram of the block is shown in Figure 5.10. Three forces act on the block: the two tension forces $(T_A$ and $T_B)$ and gravity (mg). The tensions act in two dimensions and gravity acts in the vertical direction.

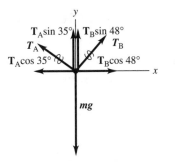

Figure 5.11 Problem 1 free-body diagram.

We have added an *xy* coordinate system to the figure to illustrate in what directions the forces act. We've also resolved the two tension forces into their *x* and *y* components.

EXECUTE We apply the equilibrium conditions to the block, writing separate equations for the *x* and *y* components:

$$\sum F_x = 0, \quad T_B \cos 48° + (-T_A \cos 35°) = 0,$$
$$\sum F_y = 0, \quad T_B \sin 48° + T_A \sin 35° + (-mg) = 0.$$

Note that components directed to the left and downward are negative, consistent with our coordinate system. We can rewrite the first equation as

$$T_B = T_A \frac{\cos 35°}{\cos 48°}.$$

Substituting for T_B in the second equation

$$T_A \frac{\cos 35°}{\cos 48°} \sin 48° + T_A \sin 35° - mg = 0$$

Solving for T_A yields

$$T_A (\cos 35° \tan 48° + \sin 35°) = mg,$$

$$T_A = \frac{mg}{(\cos 35° \tan 48° + \sin 35°)} = \frac{(322 \text{ kg})(9.80 \text{ m/s}^2)}{(\cos 35° \tan 48° + \sin 35°)} = 2130 \text{ N}.$$

Substituting the value for T_A into the first equation gives

$$T_B = T_A \frac{\cos 35°}{\cos 48°} = 2130 \text{ N} \frac{\cos 35°}{\cos 48°} = 2600 \text{ N}.$$

The tension in cable A is 2130 N and the tension in cable B is 2600 N.

EVALUATE The sum of the magnitudes of the two tension forces (4730 N) is larger than the weight of the block (3160 N). This is consistent because the tension forces are in two dimensions and their magnitudes are greater than their components. In addition, we check that the two horizontal components of the tension forces are equal in magnitude. Substituting into the terms in the first equation gives a magnitude of 1740 N for each component, each having opposite signs.

2: Accelerated motion of a block on an inclined plane

A block with mass 3.00 kg is placed on a frictionless inclined plane inclined at 35.0° above the horizontal and is connected to a second hanging block with mass 7.50 kg by a cord passing over a small, frictionless pulley (See Figure 5.11). Find the acceleration (magnitude and direction) of the 3.00 kg block.

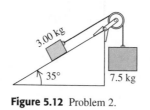

Figure 5.12 Problem 2.

Solution

IDENTIFY The blocks are accelerating, so we apply Newton's second law to both blocks. The target variable is the acceleration of the blocks (a).

SET UP Both blocks accelerate, so we'll apply Newton's second law to each block to find two equations and solve these equations simultaneously to determine the acceleration. The free-body diagrams of both blocks are shown in Figure 5.12.

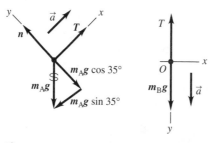

Figure 5.13 Problem 2 free-body diagrams.

The forces are identified by their magnitudes. Acting on the left-hand block (block A) are gravity ($m_A g$), the normal force (n), and the tension force (T). The right-hand block (block B) is acted upon only by the tension force (T) and gravity ($m_B g$). The tension forces must be equal in magnitude and the accelerations must be equal in magnitude, since the cord connects the two blocks. (See Conceptual Question 3.) As block B accelerates downward, block A accelerates up the ramp. We've added an xy coordinate system separately to each free-body diagram, with the positive axes aligned with the direction of acceleration. Using two different coordinate systems is preferred in these situations. The coordinate system for the right-hand block is rotated to coincide with the inclined plane. A rotated axis simplifies the analysis for ramp problems. This rotated axis requires resolving the gravity force into two components, one parallel, and one perpendicular, to the incline.

EXECUTE We apply Newton's second law to each block. Block A (with mass m_A) accelerates in the x direction (along the ramp), so

$$\sum F_x = T + \left(-m_A g \sin 35°\right) = m_A a.$$

Block B (with mass m_B) accelerates at the same rate in the y direction; hence,

$$\sum F_y = m_B g + \left(-T\right) = m_B a.$$

Both equations include the tension force, so we solve for the tension force in the second equation and substitute into the first. Our second equation becomes

$$T = m_B g - m_B a.$$

Replacing the tension force in the first equation yields

$$(m_B g - m_B a) + (-m_A g \sin 35°) = m_A a,$$

Solving for the acceleration gives

$$m_B g + (-m_A g \sin 35°) = (m_A + m_B) a,$$

$$a = \frac{g(m_B - m_A \sin 35°)}{(m_A + m_B)} = \frac{(9.80 \text{ m/s}^2)((7.50 \text{ kg}) - (3.00 \text{ kg}) \sin 35°)}{((7.50 \text{ kg}) + (3.00 \text{ kg}))} = 5.39 \text{ m/s}^2.$$

The 3.00 kg block accelerates up the ramp at 5.39 m/s². The positive value of acceleration confirms the block's acceleration up the ramp.

EVALUATE The value of acceleration is less than g, consistent with expectations. If the cord were cut, block B would accelerate at g. When block B is connected to block A through the cord, block B accelerates with an acceleration less than g. We say that block B has *additional inertia* when connected to block A.

What would have happened if we chose the direction of acceleration incorrectly? We would have found a negative acceleration, indicating that the acceleration was down the incline. In this problem, the forces do not depend on the direction of motion, and a negative acceleration would not indicate an error. It does, however, serve as a checkpoint for our calculation: A negative result with our choice of axes would cause suspicion because the right mass is larger and we expect it to accelerate downward.

Practice Problem: What mass must block A have for the system to remain at rest? Will that mass simply be 7.50 kg? *Answer:* $m_1 = 13.1$ kg, no.

3: Frictional force on an accelerating block

Two blocks are connected to each other by a light cord passing over a small, frictionless pulley as shown in Figure 5.13. Block A has mass 5.00 kg and block B has mass 4.00 kg. If block B descends at a constant acceleration of 2.00 m/s² when set in motion, find the coefficient of kinetic friction between block A and the table.

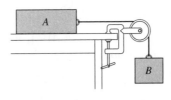

Figure 5.14 Problem 3.

Solution

IDENTIFY The target variable is the coefficient of kinetic friction. The blocks are accelerating, so we apply Newton's second law to both blocks. We'll find the friction force and determine the coefficient from that.

SET UP Both blocks accelerate, so we'll apply Newton's second law to each block to find two equations and solve those equations simultaneously. The free-body diagrams of the two blocks are shown in Figure 5.14.

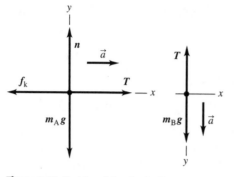

Figure 5.15 Problem 3 free-body diagram.

The forces are identified by their magnitudes. For block A, there is kinetic friction (f_k), gravity $(m_A g)$, the normal force (n), and the tension force (T). For block B there is the tension force (T) and gravity $(m_B g)$. The tension forces are equal in magnitude and the magnitudes of the acceleration are equal, as we have seen. As block B accelerates downward, block A accelerates to the right. To ensure that the acceleration of each block is in the positive direction, we added an xy coordinate system separately to each free-body diagram, with the positive axes aligned with the direction of acceleration. All forces act along the coordinate axes, so we will not need to break the forces into components.

EXECUTE We now apply Newton's second law to each block to find the friction force. Block A (with mass m_A) accelerates in the x direction, so

$$\sum F_x = T + (-f_k) = m_A a.$$

Block B (with mass m_B) accelerates at the same rate in the y direction; thus,

$$\sum F_y = m_B g + (-T) = m_B a.$$

Both equations include the tension force, so we solve for the tension force in the second equation and substitute into the first. Our second equation is then

$$T = m_B g - m_B a = m_B (g - a).$$

Replacing the tension force in the first equation gives

$$m_B (g - a) + (-f_k) = m_A a.$$

Solving for the friction force yields

$$f_k = m_B (g - a) - m_1 a = (4.00\,\text{kg})(9.80\,\text{m/s}^2 - 2.00\,\text{m/s}^2) - (5.00\,\text{kg})(2.00\,\text{m/s}^2) = 21.2\,\text{N}.$$

The friction force is related to the coefficient of kinetic friction through the normal force. We find the normal force by examining the vertical components of the forces acting on block A. There is no acceleration in the vertical direction for block A, so we can apply the equilibrium condition to block A:

$$\sum F_y = n - m_A g = 0, \qquad n = m_A g.$$

Since there are no other vertical forces acting on block A, the normal force equals the weight of block A. The kinetic frictional force is given by

$$f_k = \mu_k n,$$

which we can solve for μ_k:

$$\mu_k = \frac{f_k}{n} = \frac{f_k}{m_A g} = \frac{(21.2\,\text{N})}{(5.00\,\text{kg})(9.80\,\text{m/s}^2)} = 0.43.$$

We find the coefficient of kinetic friction between the block and the table to be 0.43.

EVALUATE A coefficient of kinetic friction equal to 0.43 compares reasonably well with values we've seen previously for smooth surfaces. Note that the tension is not equal to the weight of block B. That it is is a common misconception arising from examining the free-body diagram for block B without realizing that the block is accelerating. If we look at the rearranged second equation, the relation between the tension and the weight of block B becomes clearer:

$$T = m_B(g - a).$$

The tension force is equal to the weight only when block B's acceleration is zero. Calculating the tension for this problem, we obtain a value of 31.2 N, 20% less than the weight of block B (39.2 N).

4: Motion of a crate up a *rough* inclined plane at constant velocity

A student pushes a crate up a rough inclined plane as shown in Figure 5.15. Find the magnitude of the horizontal force the student must apply for the crate to move up the incline at constant velocity. The crate has a mass of 15.0 kg, the incline is sloped at 30.0°, and the coefficient of kinetic friction between the crate and the incline is 0.600.

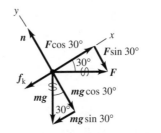

Figure 5.16 Problem 4.

Solution

IDENTIFY There is no acceleration, so we apply the equilibrium condition to the crate to find the applied force.

SET UP The free-body diagram of the crate is shown in Figure 5.16. The forces are identified by their magnitudes: kinetic frictional force (f_k), gravity (mg), the normal force (n), and the applied force (F). Kinetic friction opposes the motion up the incline and thus is directed down the incline. The rotated xy coordinate system is indicated in the diagram. This rotated axis requires resolving the gravity and applied forces into components parallel and perpendicular to the incline, as shown in the diagram.

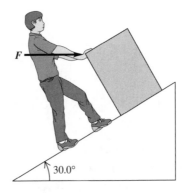

Figure 5.17 Problem 4 free-body diagram.

EXECUTE We apply the equilibrium condition to the crate. In the x-direction, along the incline, we have

$$\sum F_x = F\cos 30° + (-f_k) + (-mg\sin 30°) = 0.$$

We must apply the equilibrium condition in the y-direction to find the normal force in order to quantify the friction force. Thus,

$$\sum F_y = n + (-mg\cos 30°) + (-F\sin 30°) = 0$$
$$n = mg\cos 30° + F\sin 30°.$$

The kinetic friction is then

$$f_k = \mu_k n = \mu_k mg\cos 30° + \mu_k F\sin 30°.$$

We now substitute this result into the first equation:

$$F\cos 30° + (-\mu_k mg\cos 30° - \mu_k F\sin 30°) + (-mg\sin 30°) = 0.$$

Solving for the applied force gives

$$F\cos 30° - \mu_k F\sin 30° = \mu_k mg\cos 30° + mg\sin 30°$$
$$F(\cos 30° - \mu_k\sin 30°) = mg(\mu_k\cos 30° + \sin 30°)$$
$$F = \frac{mg(\mu_k\cos 30° + \sin 30°)}{(\cos 30° - \mu_k\sin 30°)} = \frac{(15.0\text{ kg})(9.80\text{ m/s}^2)((0.600)\cos 30° + \sin 30°)}{(\cos 30° - (0.600)\sin 30°)} = 265\text{ N}.$$

The student must push with a horizontal force of 265 N to move the crate up the incline at constant velocity.

EVALUATE A force of 265 N is roughly equivalent to the weight of a 27 kg object. Would it be easier to push the crate up the incline by pushing parallel to the incline? Yes, it would be easier to push along the ramp. In this problem, the component of the applied force directed into the incline $(F\sin 30°)$ does nothing to move the crate up the ramp. In fact, this component increases the normal force and therefore the friction force.

Practice Problem: If the student pushed along the incline, the problem would be simplified. What force would be necessary along the incline to maintain the crate at constant velocity? *Answer:* 150 N, 56% of the required horizontal force.

5: Two blocks suspended by a pulley

Two blocks are connected by a rope that passes over a small, frictionless pulley as shown in Figure 5.17. Find the tension in the rope and the acceleration of the blocks. Block 1 has a mass of 15.0 kg and block 2 has a mass of 8.0 kg.

Figure 5.18 Problem 5.

Solution

IDENTIFY Our target variables are the tension (T) and acceleration (a). We will apply Newton's second law to the problem to find the tension and acceleration.

SET UP The two blocks are separate objects, so we draw free-body diagrams of each block, shown in Figure 5.18. Two forces act on each block: gravity (mg) and the tension force (T). There is no friction in the pulley and the string is considered to be massless, so the tension in the rope is the same throughout.

We assume that the rope doesn't stretch, so the magnitudes of the two accelerations are the same, but the directions are opposite. Included in the diagrams are separate xy coordinate axes with the positive y-axis in the direction of the acceleration for both blocks (upward for block 2 and downward for block 1). This choice of axes simplifies our analysis. All of the forces act along the y-axes, so we will not need to break the forces into components.

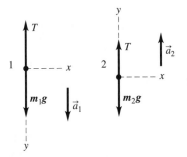

Figure 5.19 Problem 5 free-body diagrams.

EXECUTE Applying Newton's second law to both blocks gives

$$\sum F_y = m_1 g - T = m_1 a \qquad \text{(Block 1)},$$

$$\sum F_y = T - m_2 g = m_2 a \qquad \text{(Block 2)}.$$

We add both equations to eliminate T, leaving

$$m_1 g - m_2 g = m_1 a + m_2 a$$

$$a = \frac{(m_1 - m_2)g}{m_1 + m_2} = \frac{((15.0\ \text{kg}) - (8.0\ \text{kg}))(9.8\ \text{m/s}^2)}{(15.0\ \text{kg}) + (8.0\ \text{kg})} = 2.98\ \text{m/s}^2.$$

We find the tension by substituting into the second-law equation for block 2, giving

$$T = m_2(g + a) = (8.0\ \text{kg})(9.8\ \text{m/s}^2 + 2.98\ \text{m/s}^2) = 102.2\ \text{N}.$$

Both blocks accelerate at $2.98\ \text{m/s}^2$, block 1 downward and block 2 accelerates upwards. The tension in the rope is 102.2 N.

EVALUATE We check that we get the same value for the tension by using the second-law equation for block 1. We find that the equation gives a tension of 102 N, so the result is the same. We also see that the acceleration is less than $9.8\ \text{m/s}^2$, as is expected, since the net force on either block is less than its weight.

6: Friction force between two boxes

Two boxes, one on top of the other, are being pulled up a ramp at constant speed by an applied force, as shown in Figure 5.19. The coefficient of kinetic friction between box A and the ramp is 0.35, and the coefficient of static friction between the two boxes is 0.80. Box A has a mass of 3.00 kg and box B has a mass of 8.00 kg. What is the applied force?

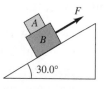

Figure 5.20 Problem 6.

Solution

IDENTIFY Our target variable is the applied force F. We will use Newton's first law to find the forces acting on the two boxes and then solve for F.

SET UP The two boxes are separate objects, so we draw free-body diagrams of each box, shown in Figure 5.20. Three forces act on box A: gravity $(m_A g)$, the normal force due to box B $(n_{B \text{ on } A})$, and static friction (f_s). Static friction must point up the ramp in order for the net force on box A to be zero. Six forces act on box B: the applied force (F), gravity $(m_B g)$, the normal force due to box A $(n_{A \text{ on } B})$, the normal force due to the ramp (n_{ramp}), kinetic friction (f_k), and static friction (f_s). Static friction and the normal force due to box A are action–reaction pairs; we set their directions opposite those of the forces acting on box A.

Included in the diagrams are a separate xy coordinate axes with the positive y-axis in the direction of the motion of both boxes (up along the ramp). We will need to break the forces into components to solve the problem.

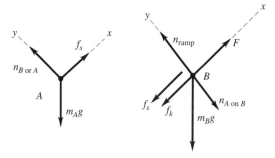

Figure 5.21 Problem 6 free-body diagrams.

EXECUTE Applying Newton's first law along both axes to both boxes gives

$$\sum F_x = f_s - m_A g \sin 30.0° = 0 \qquad \text{(box } A\text{)}$$
$$\sum F_y = n_{B \text{ on } A} - m_A g \cos 30.0° = 0 \qquad \text{(box } A\text{)}$$
$$\sum F_x = F - f_k - f_s - m_B g \sin 30.0° = 0 \qquad \text{(box } B\text{)}$$
$$\sum F_y = n_{\text{ramp}} - n_{A \text{ on } B} - m_B g \cos 30.0° = 0 \qquad \text{(box } B\text{)}.$$

We have four equations and five unknowns. To solve these equations we need the magnitude of the kinetic friction, which is

$$f_k = \mu_k n_{\text{ramp}}.$$

We begin by solving the first two equations for f_s and $n_{B \text{ on } A}$,

$$f_s = m_A g \sin 30.0° = (3.0 \text{ kg})(9.8 \text{ m/s}^2) \sin 30.0° = 14.7 \text{ N}$$
$$n_{B \text{ on } A} = m_A g \cos 30.0° = (3.0 \text{ kg})(9.8 \text{ m/s}^2) \cos 30.0° = 25.5 \text{ N}.$$

Next, we use the equation for the net force along the y-axis for box B to solve for the normal force due to the ramp, giving

$$n_{\text{ramp}} = n_{A \text{ on } B} + m_B g \cos 30.0° = (25.5 \text{ N}) + (8.0 \text{ kg})(9.8 \text{m/s}^2) \cos 30.0° = 93.4 \text{ N}.$$

We can finally find the applied force by using the equation for the net force along the x-axis for box B, along with the kinetic friction. This gives

$$F = f_k + f_s + m_B g \sin 30.0°$$
$$= \mu_k n_{\text{ramp}} + f_s + m_B g \sin 30.0°$$
$$= (0.35)(93.4 \text{ N}) + (14.7 \text{ N}) + (8.0 \text{ kg})(9.8 \text{ m/s}^2) \sin 30.0° = 86.6 \text{ N}.$$

The applied force is 86.6 N.

EVALUATE Is the coefficient of static friction large enough to keep box A from sliding off box B? We can find out by computing the maximum static friction

$$f_k^{\max} = \mu_s n_{B \text{ on } A} = (0.80)(25.5 \text{ N}) = 20.4 \text{ N}.$$

We see that the applied static friction force (14.7 N) is less than the maximum static friction force (20.4 N), so the box remains in place.

We saw how Newton's third-law force pairs can be useful in identifying forces and their directions. In this problem, we used those pairs to find the directions of forces on box B. There are also cases in which the force pairs help us identify missing force pairs. For example, the normal force acting on box B due to box A is often omitted in solutions. If we omitted $n_{A \text{ on } B}$, but had labeled the force pairs carefully, we would have realized that there was a missing normal force.

7: Acceleration in a two-pulley system

A mass is attached to a rope that is connected to the ceiling and passes through two light, frictionless pulleys. One pulley is attached to the ceiling and a second mass is attached to the other pulley, as shown in Figure 5.21. Mass 1 is 5.0 kg and mass 2 is 20.0 kg. Find the acceleration of each mass.

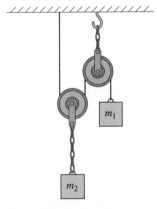

Figure 5.22 Problem 7.

Solution

IDENTIFY Our target variables are the two accelerations, a_1 and a_2, of mass 1 and mass 2, respectively. We will apply Newton's second law to find these accelerations.

SET UP The free-body diagrams for each mass are shown in Figure 5.22. Two forces act on mass 1: gravity $(m_1 g)$ and the tension in the rope (T). Three forces act on mass 2: gravity $(m_1 g)$ and the tension in the rope on both sides of the pulley (two factors of T).

Included in the diagrams are xy coordinate axes with the positive y-axis in the direction of motion for each mass (upward for mass 2 and downward for mass 1). Also shown are the accelerations a_1 and a_2 of the two masses.

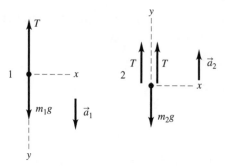

Figure 5.23 Problem 7 free-body diagrams.

EXECUTE Applying Newton's second law to each mass gives

$$\sum F_y = m_1g - T = m_1a_1 \qquad \text{(mass 1)}$$

$$\sum F_y = T + T - m_2g = m_2a_2 \qquad \text{(mass 2)}.$$

We have three unknowns in these two equations; thus, we will need more information to solve the problem. Let's examine the accelerations. As mass 1 moves down a distance L, mass 2 moves up a distance $L/2$. The change in position of mass 1 is twice the change in position of mass 2, so the velocity and acceleration of mass 1 are twice the velocity and acceleration of mass 2; therefore,

$$a_1 = 2a_2.$$

We now have enough information to solve the system of equations. We begin by replacing a_1 by a_2:

$$m_1g - T = m_12a_2$$

$$2T - m_2g = m_2a_2.$$

Doubling the first equation and adding the two equations together gives

$$2m_1g - m_2g = 4m_1a_2 + m_2a_2$$

$$a_2 = \frac{(2m_1 - m_2)g}{4m_1 + m_2} = \frac{(2(20.0 \text{ kg}) - (5.0 \text{ kg}))(9.8 \text{ m/s}^2)}{4(20.0 \text{ kg}) + (5.0 \text{ kg})} = 4.04 \text{ m/s}^2,$$

from which we obtain

$$a_1 = 2a_2 = 2(4.04 \text{ m/s}^2) = 8.08 \text{ m/s}^2.$$

Mass 1 accelerates downward at 8.08 m/s² and mass 2 accelerates upward at 4.04 m/s².

EVALUATE This problem illustrates the fact that not all accelerations are equal when objects are connected by ropes. We must always evaluate the situation carefully to determine the relation between the accelerations.

8: Coefficient of friction in a banked curve

A circular section of road with a radius of 150 m is banked at an angle of 12°. What should be the minimum coefficient of friction between the tires and the road if the roadway is designed for a speed of 25 m/s?

Solution

IDENTIFY Our target variable is the coefficient of static friction, μ_s. We will use Newton's second law to find μ_s by finding the friction force.

SET UP Figure 5.23 is a free-body diagram of the car tire on the road, showing the three forces acting on the tire: the normal force due to the road (n), static friction with the road (f), and gravity (mg). For the tire not to slip, the vertical forces must be in equilibrium and there must be a net horizontal force toward the center.

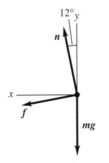

Figure 5.24 Problem 8 free-body diagram.

We have added an xy coordinate system to the figure, since the forces act in two dimensions. Note that we aligned the axes horizontally and vertically to coincide with the directions of the net forces.

SOLVE We apply the force equations to the tire, writing separate equations for the x and y components. In the vertical direction, there is no net force:

$$\sum F_y = 0, \qquad n\cos 12° - f\sin 12° - mg = 0.$$

In the horizontal direction, we use Newton's second law with centripetal acceleration:

$$\sum F_x = ma_{\text{rad}}, \qquad n\sin 12° + f\cos 12° = m\frac{v^2}{r}.$$

Note that the downward components are negative, consistent with our coordinate system. The static friction force can be replaced with $\mu_s n$ in our two equations:

$$n\cos 12° - \mu_s n\sin 12° - mg = 0,$$

$$n\sin 12° + \mu_s n\cos 12° = m\frac{v^2}{r}.$$

We now rewrite the first equation in terms of n and substitute into the second equation:

$$n = \frac{mg}{\cos 12° - \mu_s\sin 12°},$$

$$\frac{mg}{\cos 12° - \mu_s\sin 12°}\sin 12° + \mu_s\frac{mg}{\cos 12° - \mu_s\sin 12°}\cos 12° = m\frac{v^2}{r}.$$

The mass cancels and we can solve for μ_s:

$$\mu_s = \frac{v^2\cos 12° - rg\sin 12°}{v^2\sin 12° + rg\cos 12°} = \frac{(25 \text{ m/s})^2\cos 12° - (150 \text{ m})(9.8 \text{ m/s}^2)\sin 12°}{(25 \text{ m/s})^2\sin 12° + (150 \text{ m})(9.8 \text{ m/s}^2)\cos 12°} = 0.20.$$

The minimum coefficient of static friction between the tire and road is 0.20.

EVALUATE The technique for solving this problem is similar to those set forth earlier in this chapter. The differences here were the inclusion of centripetal acceleration and the choice of axes that corresponded to our knowledge of the net forces.

Practice Problem: What speed would require no frictional force? *Answer:* $v = 18$ m/s.

9: Normal force on a roller coaster

A roller coaster has a vertical loop of radius 45 m. If the roller coaster operates at a constant speed of 35 m/s while in the loop, what normal force does the seat provide for a 75 kg passenger at the top of the loop? The roller coaster is upside down at the top of the loop.

Solution

IDENTIFY Our target variable is the normal force n. We will apply Newton's second law to find the force.

SET UP Figure 5.24 shows a free-body diagram of the passenger on the roller coaster. The figure shows the two forces acting on the passenger: the normal force due to the seat (n) and gravity (mg). At the top of the loop, the net force is downward and the person is accelerating toward the center. We have added an xy coordinate system to the figure, with positive forces directed downward.

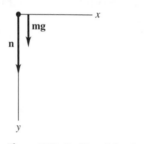

Figure 5.25 Problem 9 free-body diagram

SOLVE The net force on the passenger is directed downward. Newton's second law with centripetal acceleration gives

$$\sum F_y = ma_{rad}, \qquad n + mg = m\frac{v^2}{r}.$$

Solving for the normal force, we obtain

$$n = m\frac{v^2}{r} - mg = (75\text{ kg})\frac{(35\text{ m/s})^2}{(45\text{ m})} - (75\text{ kg})(9.8\text{ m/s}^2) = 1300\text{ N}.$$

The seat exerts a force of 1300 N on the passenger.

REFLECT We see that the seat provides a force nearly twice the passenger's weight. A seat belt would not be needed to prevent a fall from the roller coaster at the top of the loop. Amusement park rides get their reputation for excitement from their ability to rapidly change the magnitudes and directions of forces applied to passengers. The normal force of the seat on the passenger is even larger at the bottom of the loop.

Practice Problem: What normal force does the seat provide at the bottom of the loop? *Answer:* $N = 2800$ N.

10: Investigating a tetherball

A tetherball is attached to a vertical pole with a 2.0 m length of rope, as shown in Figure 5.25. If the rope makes an angle of 25.0° with the vertical pole, find the time required for one revolution of the tetherball.

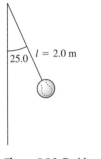

Figure 5.26 Problem 10.

Solution

IDENTIFY Our target variable is the period of one revolution, T. The period will be found from the ball's velocity and circumference. The velocity will be found by applying Newton's second law to the ball as it undergoes centripetal acceleration.

SET UP The free-body diagram of the tetherball is shown in Figure 5.26. Two forces act on the ball: gravity (mg) and the tension in the rope (T). An xy coordinate axis is included in the diagram.

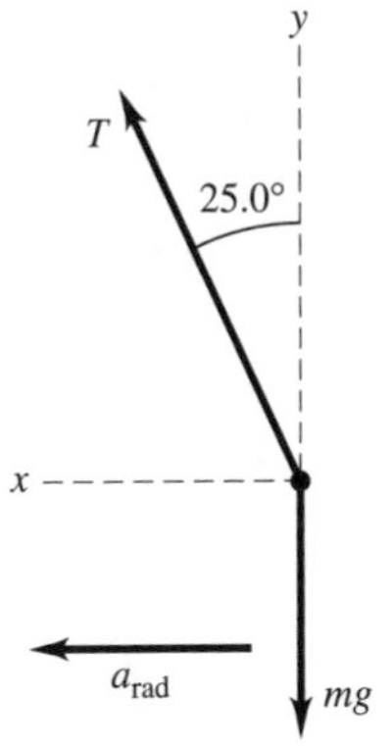

Figure 5.27 Problem 10 free-body diagram.

EXECUTE The tetherball undergoes centripetal acceleration in the horizontal direction and no net force in the vertical direction. Applying Newton's first law in the vertical direction gives

$$\sum F_y = T\cos 25.0° - mg = 0$$

$$T = \frac{mg}{\cos 25.0°}.$$

Applying Newton's second law in the horizontal direction gives

$$\sum F_x = T\sin 25.0° = ma_{\text{rad}} = m\frac{v^2}{R}$$

$$v^2 = T\frac{R}{m}\sin 25.0° = \frac{mg}{\cos 25.0°}\frac{R}{m}\sin 25.0° = Rg\tan 25.0°.$$

We now have the velocity in terms of the radius, angle, and g. To find the radius, we use the length of the rope and the sine:

$$R = l \sin 25.0° = (2.0 \text{ m}) \sin 25.0° = 0.845 \text{ m}.$$

Thus,

$$v = \sqrt{Rg \tan 25.0°} = \sqrt{(0.845 \text{ m})(9.8 \text{ m/s}^2) \tan 25.0°} = 1.97 \text{ m/s}.$$

The period is the time required for one revolution. The ball travels the circumference of a circle of radius R in one period. The period is then

$$T = \frac{2\pi R}{v} = \frac{2\pi(0.845 \text{ m})}{(1.97 \text{ m/s})} = 2.69 \text{ s}.$$

The ball completes one revolution in 2.69 s.

EVALUATE This problem illustrates the inclusion of centripetal acceleration into force problems. We see that Newton's laws remain unchanged. We simply set the acceleration equal to the radial acceleration in these problems.

Practice Problem: Does the period increase or decrease with larger angles? Find the period when the rope makes an angle of 50.0° with respect to the vertical. *Answer:* 2.28 s; the period decreased slightly.

11: A rotating mass

A mass attached to a vertical post by two strings rotates in a circle of constant velocity v. (See Figure 5.27.) At high velocities, both strings are taut. Below a critical velocity, the lower string slackens. Find the critical velocity. The mass is 1.0 kg and is 1.5 m from the post.

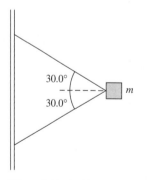

Figure 5.28 Problem 11.

Solution

IDENTIFY Our target variable is the critical velocity, at which the tension in the lower string is zero. We will find the tension in the lower string and the velocity by applying Newton's second law. Then we'll set the tension in the lower string to zero and solve for the critical velocity.

SET UP The free-body diagram of the mass is shown in Figure 5.28. Three forces act on the mass: gravity (mg), the tension in the upper rope (T_1), and the tension in the lower rope (T_2). An xy coordinate axis is included in the diagram.

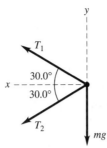

Figure 5.29 Problem 11 free-body diagram.

EXECUTE The mass undergoes centripetal acceleration in the horizontal direction and no net force in the vertical direction. Applying Newton's first law in the vertical direction gives

$$\sum F_y = T_1 \sin 30° - T_2 \sin 30° - mg = 0.$$

Applying Newton's second law in the horizontal direction gives

$$\sum F_x = T_1 \cos 30° + T_2 \cos 30° = ma_{\text{rad}} = m\frac{v^2}{R}.$$

We can solve for T1 by multiplying the first equation by cos 30°, multiplying the second equation by sin 30°, and adding the two resulting equations. Doing this gives

$$2T_1 \sin 30° \cos 30° = mg \cos 30° + m\frac{v^2}{R} \sin 30°,$$

which reduces to

$$T_1 = \frac{mg}{2\sin 30°} + \frac{mv^2}{2R\cos 30°}.$$

Solving for T_2 produces

$$T_2 = \frac{-mg}{2\sin 30°} + \frac{mv^2}{2R\cos 30°}.$$

At the critical velocity, T_2 is zero. Solving for v, we obtain

$$\frac{mg}{2\sin 30°} = \frac{mv^2}{2R\cos 30°},$$

$$v = \sqrt{\frac{gR}{\tan 30°}} = \sqrt{\frac{(9.8 \text{ m/s}^2)(1.5 \text{ m})}{\tan 30°}} = 5.04 \text{ m/s}.$$

The critical velocity is 504 m/s.

EVALUATE For velocities less than the critical velocity, the bottom string is slack. Strings always have positive tensions. When they are slack, they provide no support.

Practice Problem: What is the tension in the upper string at the critical velocity? *Answer:* 20 N.

Try It Yourself!

1: Constant velocity on an incline

Two weights are attached by a light cord that passes over a light, frictionless pulley as shown in Figure 5.29. The left weight moves up a rough ramp. Find the weight w_2 necessary to keep w_1 (15.0 N) moving up the ramp at a constant rate once it is put in motion. The coefficient of kinetic friction is 0.25 and the ramp is inclined at 30.0°.

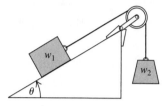

Figure 5.30 Try it yourself 1.

Solution Checkpoints

IDENTIFY AND SET UP To move at constant velocity, the net force on each weight must be zero. The rough surface indicates that there is friction between the weight and the ramp. Draw a free-body diagram and apply Newton's first law to solve.

EXECUTE The net force acting on weight 1 along the ramp is

$$\sum F_x = T - f - w_1 \sin\theta = 0.$$

The net force acting on weight 2 in the vertical direction is

$$\sum F_y = T - w_2 = 0.$$

These two equations have three unknowns. You need to find another expression to solve for w_2. The expression will lead to

$$w_2 = \mu_k w_1 \cos\theta + w_1 \sin\theta = 11.1 \text{ N}$$

EVALUATE Can you explain why w_2 has less weight than w_1?

2: Box sliding across rough floor

A box is kicked, giving it an initial velocity of 2.0 m/s. It slides across a rough, horizontal floor and comes to rest 1.0 m from its initial position. Find the coefficient of friction.

Solution Checkpoints

IDENTIFY AND SET UP Begin with a sketch and free-body diagram. In the horizontal direction, the only force acting on the box is friction. Use the acceleration of the box and Newton's second law to find the friction.

EXECUTE The net horizontal force acting on the box is

$$\sum F_x = f = ma.$$

To find the acceleration, apply

$$v^2 = v_0^2 + 2a\Delta x,$$

one of the kinematics relations for constant acceleration that we've used in previous chapters. You should find that the coefficient of kinetic friction is 0.20.

EVALUATE We see how we can combine kinematics with our force problems in this problem. Did we find a reasonable coefficient of kinetic friction?

3: Three connected masses

Three blocks are attached to each other with ropes that pass over pulleys as shown in Figure 5.30. The masses of the ropes and pulleys can be ignored, and there is no friction on the surface over which the blocks slide or in the pulleys. Find the tension in the two ropes and the acceleration of the blocks.

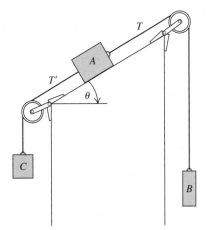

Figure 5.31 Try it yourself 2.

Solution Checkpoints

IDENTIFY AND SET UP Three objects are in motion, so you must draw three free-body diagrams. How does the acceleration of all three objects compare? You can apply Newton's laws to solve. Assume that the system accelerates clockwise.

EXECUTE The net forces acting in the direction of the acceleration of the blocks are

$$\sum F_A = T - T' - m_A g \sin\theta = m_A a$$
$$\sum F_B = m_B g - T = m_B a$$
$$\sum F_C = T' - m_C g = m_C a.$$

We can also find the net force acting on block A normal to the ramp. Using these equations to solve for the tensions gives

$$T = \left(\frac{2 m_C m_B + m_B m_A (1 + \sin\theta)}{m_A + m_B + m_C} \right) g$$

$$T' = \left(\frac{2 m_C m_B + m_C m_A (1 - \sin\theta)}{m_A + m_B + m_C} \right) g.$$

Solving for the acceleration yields

$$a = \left(\frac{m_B - m_A \sin\theta - m_C}{m_A + m_B + m_C} \right) g.$$

EVALUATE How do we interpret the results if we find that the acceleration is negative? Do we need to rework the solution? Is the acceleration greater or less than g?

4: Tension along a heavy rope

A heavy rope of mass 10.0 kg and length 5.0 m lies on a frictionless horizontal surface. If a certain horizontal force is applied, the rope accelerates at 1.5 m/s^2. Find the tension in the rope at any point along its length.

Solution Checkpoints

IDENTIFY AND SET UP To find the tension at any point in the rope, you must break the rope into many pieces and find the tension of any piece. It is easiest to pick a piece of the rope at the end where the force is applied, as shown in Figure 5.31. By applying Newton's second law to this piece, you can find the tension anywhere in the rope.

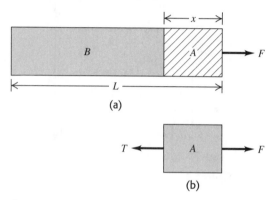

Figure 5.32 Try it yourself 3.

EXECUTE The net force required to accelerate the chain is 15.0 N. The mass of the length x of the rope is

$$m_x = \frac{x}{L}m.$$

The net force on the piece of rope is

$$\sum F_x = F - T = m_x a = \frac{x}{L}ma.$$

Solving for T to find the tension as a function of x:

$$T = F\left(1 - \frac{x}{L}\right) = (10.0 \text{ N})\left(1 - \frac{x}{5.0 \text{ m}}\right).$$

EVALUATE This example illustrates how to apply Newton's second law to more complicated problems. Does the tension in the rope increase or decrease as you move away from the end where the force is applied?

5: Riding a carousel

A 100.0 kg man stands on the outer edge of a carousel of 4.0 m radius. The coefficient of static friction between his shoes and the carousel is 0.30. What is the minimum period of rotation required for the man to remain on the carousel?

Solution Checkpoints

IDENTIFY AND SET UP Three forces act on the man: gravity, the normal force, and friction. The condition required for him to slip is that the centripetal force exceed the static friction. The period is determined from the velocity.

EXECUTE The net horizontal force acting on the man is

$$\sum F_x = f = ma_{\text{rad}} = m\frac{v^2}{R}.$$

This yields the critical velocity when the friction force is equal to the centripetal force. For velocities larger than the critical velocity, the man will slide off. The period is found from the velocity, given that the time required for one revolution is the distance traveled divided by the velocity. The minimum period is 7.4 s.

EVALUATE We see that the period and velocity are inversely related: Larger velocities result in shorter periods. How does the friction force vary when the period is greater than 7.4 s?

Problem Summary

Chapters 4 and 5 have examined a variety of problems with applied forces in diverse applications, but they share a common problem-solving foundation. For all problems, we

- Identified the general procedure for finding the solution.
- Sketched the situation when no figure was provided.
- Identified the forces acting on the objects of interest.
- Drew free-body diagrams of forces acting *on* the objects.
- Added appropriate coordinate systems to the free-body diagrams.
- Applied the equilibrium condition, Newton's second law, or both to the objects in order to find relations among the forces, masses, and accelerations.
- Solved the equations through algebra, trigonometry, and calculus.
- Reflected on the results, thus checking for inconsistencies.

This problem-solving foundation can be applied to all problems involving forces. Following this procedure enables one to master Newton's laws.

6 Work and Kinetic Energy

Summary

We introduce two new concepts in this chapter: *work* and *energy.* Our investigation begins by learning how work can be used to solve problems with variable forces. We will see how work is a form of energy transfer, leading us to learn about energy, one of the most important concepts in physics. We'll learn how work can be used to change a body's kinetic energy (the energy of motion), how to determine the work expended in many situations, and how power is the rate of change of work with respect to time. In the next chapter, we will introduce the law of conservation of energy and discover other forms of energy. The problem-solving skills we develop in these two chapters will prepare us for additional forms of energy that we'll encounter in later chapters.

Objectives

After studying this chapter, you will understand

- The definition of work and how to calculate the work done by a force on a body.
- The definition and interpretation of kinetic energy.
- How to apply the work-energy theorem to problems.
- How to use kinetic energy and work in problems involving varying forces applied along curved paths.
- How to analyze springs and the elastic force.
- The definition of power and how to calculate power for bodies performing work or on which work is performed.

Concepts and Equations

Term	Description
Work Done by a Force	A constant force acting on and displacing a body does work. For a constant force $\vec{F}$ acting on a particle causing a straight-line displacement $\vec{s}$ at an angle ϕ with respect to the force, the work done by the force on the body is $$W = \vec{F} \cdot \vec{s} = Fs\cos\phi.$$ The SI unit of work is 1 joule = 1 newton-meter $\left(1\text{ J} = 1\text{N} \cdot \text{m}\right)$.
Kinetic Energy	Kinetic energy K is the energy of motion of a particle with mass. Kinetic energy is equal to the amount of work required to accelerate a particle from rest to a speed v. A particle of mass m and velocity v has kinetic energy $$K = \tfrac{1}{2}mv^2.$$
Work-Energy Theorem	The work–energy theorem states that the total work done by a net external force on a particle as it undergoes a displacement is equal to the change in kinetic energy of the particle: $$W_{\text{tot}} = K_2 - K_1 = \Delta K.$$
Work Done by a Varying Force along a Curved Path	The work done by a varying force on a particle as it follows a curved path is determined by $$W = \int_{P_1}^{P_2} \vec{F} \cdot d\vec{l} = \int_{P_1}^{P_2} F\cos\phi\, dl = \int_{P_1}^{P_2} F_s dl.$$
Elastic Force	An elastic force is a force that restores a body to its original equilibrium position after deformation. For a spring, the deformation is approximately proportional to the applied force, as given by Hooke's law, $$F_{\text{spr}} = kx,$$ where k is the force constant and x is the displacement of the spring from its equilibrium position.
Power	Power is the rate of change of work with respect to time. Average power is defined as $$P_{\text{av}} = \frac{\Delta W}{\Delta t},$$ where ΔW is the quantity of work performed during the time interval Δt. Instantaneous power is defined as $$P = \lim_{\Delta t \to 0}\frac{\Delta W}{\Delta t} = \frac{dW}{dt}.$$ For a force acting on a moving particle, the instantaneous power is $$P = \vec{F} \cdot \vec{v}.$$ The SI unit of power is 1 watt = 1 joule/second $\left(1\text{ W} = 1\text{ J/s}\right)$.

Conceptual Questions

1: Work done by the normal force

How much work does the normal force do on a box sliding across the floor?

Solution

IDENTIFY, SET UP, AND EXECUTE Work is produced when a force acts on an object in the direction the object is displaced. As a box slides across the floor, the normal force is perpendicular to its motion. Therefore, the normal force does no work on the sliding box.

EVALUATE Work has a strict definition in physics. The normal force prevents the box from falling into the floor, but it does no work thereby. You also do no work as you carry your backpack, even though you arm tires.

2: Which force does the most work?

Rank the four situations shown in Figure 6.1 from most to least work done by the force. The displacement is the same in each case.

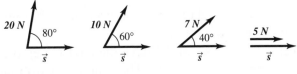

Figure 6.1 Conceptual Question 2.

Solution

IDENTIFY, SET UP, AND EXECUTE Work is the dot product of force and displacement, or displacement times the component of force parallel to the displacement. All four situations have the same displacement, so we rank the components of the force parallel to the displacement. The parallel components of the forces are the magnitudes of the forces times the cosine of the angle between the force and the displacement vectors.

We find that the parallel components for the four situations are, in order, 3.47 N, 4.00 N, 5.36 N, and 5.00 N. Therefore, the ranking from greatest to least amount of work is (c), (d), (b), and (a).

EVALUATE We see that larger forces do not necessarily produce more work. Work depends on *both* the magnitude of the force *and* the direction of the force with respect to the displacement. The 20 N force produces much less work than the 5 N force.

3: Ranking stopping distance

Five identical shipping crates slide down a ramp onto a rough horizontal floor. Each crate carries a different mass and has a different velocity at the bottom of the ramp, given in Table 1. Rank the distances required, from least to greatest, for each crate to stop.

TABLE 1: Conceptual Problem 3.

	Mass (kg)	Velocity at bottom of ramp (m/s)
Crate 1	20.0	10.0
Crate 2	30.0	20.0
Crate 3	20.0	15.0
Crate 4	100.0	10.0
Crate 5	200.0	5.0

Solution

IDENTIFY, SET UP, AND EXECUTE At the bottom of the ramp, each crate has kinetic energy. As they slow, the crates lose kinetic energy because friction does work on them. The change in kinetic energy is equal to the work done by friction, which in turn is equal to the friction force (μmg) times the displacement (x). Since the crates are identical, their coefficient of friction is the same. Algebraically, the work done is

$$W = \Delta K = K_2 - K_1,$$

$$\mu mgx = 0 - \tfrac{1}{2}mv^2,$$

$$\mu gx = -\tfrac{1}{2}v^2.$$

We see that the displacement is proportional to the velocity squared. To rank the stopping distances, we compare the velocities at the bottom of the ramp.

Crate 5 has the smallest velocity, crates 1 and 4 have the next-largest velocity, crate 3 has the next-largest velocity, and crate 2 has the largest velocity. The ranking of stopping distance for the crates is (5), (1) = (4), (3), and (2).

EVALUATE We see that in this case the results do not depend on the mass of the object. Crate 4 has much more mass than crate 1, so crate 4 has more initial kinetic energy. However, the effect of friction on crate 4 is greater than the effect of friction on crate 1. These two effects cancel, resulting in the same stopping distance.

We also see that the work done by friction is negative, because the change in kinetic energy is negative.

4: Work in carnival ride

A swing ride at a carnival consists of chairs attached by a cable to a vertical pole. The vertical pole rotates, causing the chairs to swing in a circle as shown in Figure 6.2. How much work is done by the tension in the cable as one chair (with a mass of 70 kg) makes one complete revolution? The length of the rope and angle are shown in the figure.

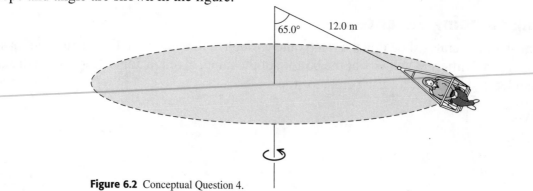

Figure 6.2 Conceptual Question 4.

Solution

IDENTIFY, SET UP, AND EXECUTE As the chair swings, it is displaced along the circle. The tension in the cable is directed perpendicular to the chair's displacement. Since work is the dot product of the force and displacement, the work done by the tension is zero.

EVALUATE Does the kinetic energy change? The chair moves at a constant rate, so the kinetic energy remains constant. No work is needed to keep the chair moving at the same speed.

5: Work with equal forces

A force is applied to several boxes. Rank the five situations shown in Figure 6.3 from least to most work done by the force. The same magnitude of force is applied to all boxes, and the mass of each box is given.

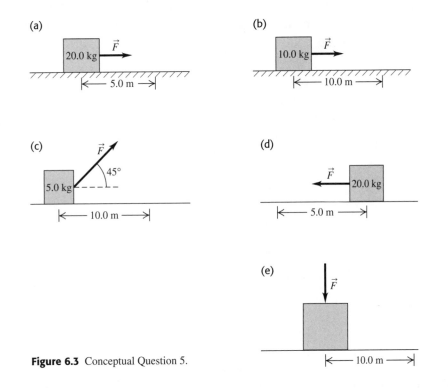

Figure 6.3 Conceptual Question 5.

Solution

IDENTIFY, SET UP, AND EXECUTE Work is the dot product of force and displacement. All five situations have the same magnitude of applied force, so we rank them on the basis of the displacement and the cosine of the angle between the force and the resulting displacement.

In cases (a) and (b), the displacement is in the direction of the force. The box in (b) is displaced further, so more work is done by the force. In case (c), the force is applied at an angle of 45° with respect to the displacement and the box is displaced 10.0 m, resulting in less work than in (b), but more than in (a). The work in case (d) is negative, since the force is opposite the displacement. In case (e), no work is done, since the force is perpendicular to the displacement.

The amount of work done by the force, from least to most, is (d), (e), (a), (c), and (b).

EVALUATE We see that the mass of the box does not influence the results: The work done by the force depends only on the force, the displacement, and the angle between them.

How much work is done on the box in (e)? The box could be moving at constant speed on a frictionless surface, in which case no work is done. Or another force may be acting on the box, creating work. From the information we have, we cannot determine which is correct.

We also see that we can have negative work in problems. Negative work indicates that energy is being removed from the system, perhaps as the box slows to a stop.

Problems

1: Work done in pushing a crate up an inclined plane.

A student pushes a crate 3.50 m up a rough inclined plane with a constant horizontal force of 225 N, starting from rest as shown in Figure 6.4. Find the work done by the student, the work done by friction, the work done by gravity, and the change in the crate's kinetic energy. How does the work done by the student, friction, and gravity compare with the change in kinetic energy? The crate has a mass of 15.0 kg, the incline is sloped at 30.0°, and the coefficient of kinetic friction between the crate and the incline is 0.400.

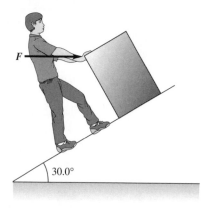

Figure 6.4 Problem 1.

Solution

IDENTIFY Each force is constant and the displacement is along a straight line, so we can find the work and kinetic energy from their definitions.

SET UP We'll begin with a free-body diagram to find the work done by the three forces. Figure 6.5 shows the free-body diagram with a rotated coordinate system that coincides with the incline to simplify the analysis. The forces are identified by their magnitudes: kinetic friction (f_k), gravity (mg), the normal force (n), and the force applied by the student (F). Kinetic friction opposes the motion up the incline and thus is directed down the incline. Gravity and the applied force are resolved into components parallel and perpendicular to the incline.

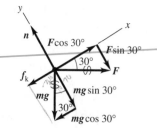

Figure 6.5 Problem 1 free-body diagram.

EXECUTE The work done by the student pushing the crate is

$$W_{student} = Fs\cos\phi = (225 \text{ N})(3.50 \text{ m})\cos 30° = 682 \text{ J},$$

where the angle between the force and the displacement is 30°. To find the work done by friction, we need to know the friction force. We apply the equilibrium condition in the y direction to find the normal force in order to quantify the friction force:

$$\sum F_y = n + (-mg\cos 30°) + (-F\sin 30°) = 0,$$

$$n = mg\cos 30° + F\sin 30° = (15.0 \text{ kg})(9.8 \text{ m/s}^2)(\cos 30°) + (225 \text{ N})(\sin 30°) = 240 \text{ N}.$$

The kinetic friction force is then

$$f_k = \mu_k n = (0.400)(240 \text{ N}) = 95.9 \text{ N}.$$

Friction is directed opposite to the displacement, so the angle between them is 180°. The work done by friction is

$$W_{f_k} = f_k s\cos\phi = (95.9 \text{ N})(3.50 \text{ m})\cos 180° = -336 \text{ J}.$$

The component of gravity along the displacement is also opposite to the displacement. The work done by gravity is

$$W_{grav} = F_g s\cos\phi = mgs\cos 120° = (15.0 \text{ kg})(9.8 \text{ m/s}^2)(3.50 \text{ m})\cos 120° = -257 \text{ J}.$$

To find the change in kinetic energy, we need the initial and final velocities. The initial velocity is zero. The final velocity is found by applying Newton's second law and kinematics. In the x-direction, along the incline, Newton's second law gives

$$\sum F_x = F\cos 30° + (-f_k) + (-mg\sin 30°) = ma.$$

The acceleration of the box is then

$$a = \frac{F\cos 30° + (-f_k) + (-mg\sin 30°)}{m}$$

$$= \frac{(225 \text{ N})\cos 30° - (95.9 \text{ N}) - (15.0 \text{ kg})(9.8 \text{ m/s}^2)\sin 30°}{(15.0 \text{ kg})} = 1.70 \text{ m/s}^2.$$

Constant-acceleration kinematics gives the final velocity:

$$v^2 = v_{x0}^2 + 2a_x(x - x_0),$$

$$v = \sqrt{0 + 2(1.70 \text{ m/s}^2)(3.50 \text{ m})} = 3.45 \text{ m/s}.$$

The change in kinetic energy is then

$$\Delta K = K_2 - K_1 = \tfrac{1}{2}mv^2 - 0 = \tfrac{1}{2}(15.0 \text{ kg})(3.45 \text{ m/s})^2 = 89.3 \text{ J}.$$

In sum, we found that the student did 682 J of work on the crate, friction did -336 J of work on the crate, gravity did -257 J of work, and the kinetic energy increased by 89 J. When we add the work due to the three forces together, the total work is 89 J. The total work is equal to the change in kinetic energy.

EVALUATE This problem affords a thorough investigation of work and kinetic energy. It illustrates how to combine the work due to several forces into the total work and how the total work on a system is used to increase the kinetic energy of the system.

CAUTION   Watch Signs! You must evaluate the signs carefully when determining work and energy. Negative work indicates that the force is directed opposite to the displacement and often slows the object, as it does in this problem.

2: Spring force between two blocks

Two blocks are placed on a horizontal, frictionless surface and attached to each other by a spring with force constant 4500 N/m. If the right-hand block is pulled with a force of 150.0 N, find the displacement of the spring as the blocks accelerate. The left-hand block has a mass of 5.00 kg, and the right-hand block has a mass of 3.00 kg.

Solution

IDENTIFY   Our target variable is the displacement of the spring. We can solve this problem with Newton's second law.

SET UP   The displacement of the spring is proportional to the spring force. We find the spring force by applying Newton's second law to the blocks and then use Hooke's law to find the displacement. The first task is to sketch the situation, as shown in Figure 6.6.

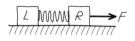

Figure 6.6 Problem 2.

Examining the sketch, we see the two blocks interact through the spring. We draw free-body diagrams for the two blocks, shown in Figure 6.7.

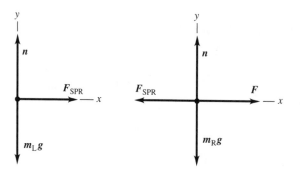

Figure 6.7 Problem 2 free-body diagram.

The forces are identified by their magnitudes: force due to the spring (F_{spr}), gravity (mg), the normal force (n), and the applied force (F). Knowledge of the vertical forces will not be necessary to solve this problem. The blocks are connected to each other; therefore, both accelerate at the same rate and are acted upon by the same magnitude of spring force. The diagrams include a common xy coordinate axis, with the positive x-axis in the direction of the acceleration. All of the forces act along the coordinate axes, so we will not need to break the forces into components.

EXECUTE   We apply Newton's second law to the horizontal forces acting on each block to determine the force due to the spring. For the right-hand block (with mass m_R),

$$\sum F_x = F + (-F_{spr}) = m_R a.$$

For the left-hand block (with mass m_L),

$$\sum F_x = F_{spr} = m_L a.$$

Examining these two equations, we find two unknowns: F_{spr} and a. We wish to find the spring force, so we rewrite the second equation in terms of acceleration:

$$a = \frac{F_{spr}}{m_L}.$$

Substituting for the acceleration in the first equation yields

$$F + (-F_{spr}) = m_R \frac{F_{spr}}{m_L},$$

$$F = F_{spr} + F_{spr} \frac{m_R}{m_L} = F_{spr}\left(1 + \frac{m_R}{m_L}\right),$$

$$F_{spr} = \frac{F}{\left(1 + \frac{m_R}{m_L}\right)} = F \frac{m_L}{m_R + m_L} = (150)\frac{(5.00\text{ kg})}{(3.00\text{ kg}) + (5.00\text{ kg})} = 93.8\text{ N}.$$

This gives us the magnitude of the force due to the spring. The direction is opposite to the displacement. We can now use Hooke's law,

$$F_{spr} = kx,$$

to solve for the displacement:

$$x = \frac{F_{spr}}{k} = \frac{(93.8\text{ N})}{4500\text{ N/m}} = 0.0208\text{ m} = 2.08\text{ cm}.$$

The spring is displaced 2.08 cm when the blocks are pulled.

EVALUATE We see that the force due to the spring is less than the applied force in this problem. That is reasonable, as the spring must provide force to accelerate only the right-hand mass, while the applied force must accelerate both masses.

3: Block stopped by spring and friction

A 5.00 kg block is moving along a rough horizontal surface toward a spring with force constant 500 N/m. The velocity of the block just before it contacts the spring is 12.0 m/s. If the coefficient of kinetic friction between the block and the surface is 0.400, what is the maximum compression of the spring?

Solution

IDENTIFY We can apply the work–energy theorem to the problem. Two forces (friction and the spring force) do work to slow the block, taking away the kinetic energy from the block. The target variable is the spring's maximum compression.

SET UP Figure 6.8 shows a sketch of the situation. Just before contacting the spring, the block has kinetic energy and the spring is uncompressed. At maximum compression, the spring has been compressed a distance X, the block is not moving ($K = 0$), and friction has done work on the block as it moves the same distance X. The work–energy theorem tells us that the change in kinetic energy is equal to the total work done.

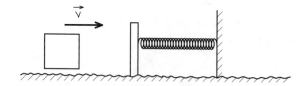

Figure 6.8 Problem 3 sketch.

Four forces act on the block as it slows: the force due to the spring (F_{spr}), gravity (mg), the normal force (n), and the force of kinetic friction (f_k). These forces are shown in the free-body diagram in Figure 6.9. The normal force and gravity do no work on the block as it slows.

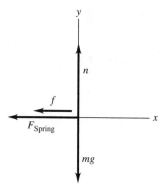

Figure 6.9 Problem 3 free-body diagram.

EXECUTE The change in kinetic energy of the block is

$$\Delta K = K_2 - K_1 = 0 - \tfrac{1}{2}mv^2 = -\tfrac{1}{2}mv^2.$$

The work done on the spring by the block is

$$W_{\text{Block on Spring}} = \tfrac{1}{2}kx_2^2 - \tfrac{1}{2}kx_1^2 = \tfrac{1}{2}kX^2.$$

The work done by the spring on the block is the negative of this value:

$$W_{\text{Spring}} = -\tfrac{1}{2}kX^2.$$

The work done by friction on the block is

$$W_f = Fs\cos\phi = (\mu mg)s\cos 180° = -\mu mgX.$$

The total work is the work due to the spring and the work due to friction. Setting the sum of these two quantities equal to the change in kinetic energy gives

$$-\tfrac{1}{2}kX^2 - \mu mgX = -\tfrac{1}{2}mv^2.$$

This is a quadratic equation. Solving for X yields

$$X = \frac{-(\mu mg) \pm \sqrt{(\mu mg)^2 - 4(\tfrac{1}{2}k)(-\tfrac{1}{2}mv^2)}}{2(\tfrac{1}{2}k)}$$

$$= \frac{-(19.6) \pm \sqrt{(19.6)^2 + 4(15.0)(22.5)}}{2(15.0)} \, m = 0.735, -2.04 \text{ m}.$$

In this case, we require the positive root, 0.735 m. The spring compresses 0.735 m when the block comes to a momentary stop.

EVALUATE This problem illustrates how to use energy to work with varying forces. The analysis would have been more difficult had we used Newton's second law, as we would then have had to integrate the force equation.

CAUTION   **Watch signs in work done by a spring!** As you pull on a spring, you do positive work on it and the spring does negative work on you. Understanding the signs for work done on or by a spring will help you understand the proper signs for work and energy.

4: A novel spring

Suppose you have invented a novel spring that will slow a 2.5 kg toy car. The spring exerts a force that depends on position such that $F_x = [10.0 \text{ N} + (16.0 \text{ N/m}^2) \, x^2]$. What maximum speed of the toy car will the spring stop with a compression of 0.50 m?

Solution

IDENTIFY The maximum work due to the spring occurs when the displacement is directed along the force, so we will take that direction as the direction of displacement. Our target variable is the speed of the toy car, and we will find it by using the work–energy theorem.

SET UP Only the force due to the novel spring acts on the toy car. Initially, there is just the kinetic energy of the car. At the maximum compression, there is no kinetic energy.

EXECUTE The force of the spring is in the direction of the motion. The work that the spring does on the toy car as the spring is compressed a distance X is

$$W = \int_0^X F_x dx.$$

Substituting for the force and finding the work when $X = 0.50$ m yields

$$W = \int_0^X \left[10.0 \text{ N} + \left(16.0 \frac{\text{N}}{\text{m}^2} \right) x^2 \right] dx$$

$$= \left[(10.0 \text{ N})x + \left(16.0 \frac{\text{N}}{\text{m}^2} \right) \frac{x^3}{3} \right] \Big|_0^X$$

$$= \left[(10.0 \text{ N})X + \left(16.0 \frac{\text{N}}{\text{m}^2} \right) \frac{X^3}{3} \right]$$

$$= \left[(10.0 \text{ N})(0.50 \text{ m}) + \left(16.0 \frac{\text{N}}{\text{m}^2} \right) \frac{(0.50 \text{ m})^3}{3} \right]$$

$$= 6.33 \text{ J}.$$

The spring can stop a toy car with up to 6.33 J of energy. The velocity of that car is determined from the kinetic energy:

$$K = \tfrac{1}{2}mv^2,$$

$$v = \sqrt{\frac{2K}{m}} = \sqrt{\frac{2(6.33\,\text{J})}{(2.5\,\text{kg})}} = 2.25\,\text{m/s}.$$

The maximum velocity that the spring can stop is 2.25 m/s.

EVALUATE In this problem, the work is negative because the force due to the spring acts opposite to the direction of the displacement in order to reduce the kinetic energy. The change in kinetic energy is also negative, so the work–energy theorem is satisfied.

5: Average power to run an escalator

What average power does an escalator require to lift twenty 100.0 kg people 3.0 m high in 1 minute?

Solution

IDENTIFY The target variable is the power, or the amount of work done per unit time.

SET UP The escalator provides work equivalent to the amount of work due to gravity as the people are lifted. We will find the work done by gravity and divide by the time.

EXECUTE The work done by gravity is the weight of the people, multiplied by their change in height:

$$W = mgh = (20 \times 100\,\text{kg})(9.8\,\text{m/s}^2)(3.0\,\text{m}) = 58{,}800\,\text{J}.$$

The power is the work of gravity divided by time:

$$P = \frac{\Delta W}{\Delta t} = \frac{(58{,}800\,\text{J})}{(60\,\text{s})} = 980\,\text{W}.$$

The power needed to run the escalator is 980 W.

EVALUATE We see that the power is independent of the angle of the escalator. This is due to the fact that the gravitational work depends only on the change in vertical elevation.

Try It Yourself!

1: Box on a smooth incline

A 10.0 kg box is pushed 2.0 m up a smooth inclined plane of angle 30.0° by a 100 N horizontal force, starting from rest. Find the work done on the box and the change in kinetic energy.

Solution Checkpoints

IDENTIFY AND SET UP Start with a free-body diagram to identify the forces acting on the box. The work due to each force is equal to the dot product of the force and the displacement. The change in kinetic energy is equal to the total work done.

EXECUTE Three forces act on the box. The work done by each is

$$W_{\text{applied}} = Fs\cos\phi = 173\,\text{J},$$
$$W_{\text{normal}} = 0,$$
$$W_g = -98\,\text{J}.$$

The change in kinetic energy is 75 J.

EVALUATE You can check your results by using Newton's second law to find the acceleration, which leads to the final velocity of the box and final kinetic energy of the box. Do they agree?

2: Body sliding on a rough surface

A body slides on a rough surface. If the body is given an initial velocity of 3.0 m/s, it comes to a stop in 1.0 m. Find the coefficient of kinetic friction between the body and the surface.

Solution Checkpoints

IDENTIFY AND SET UP Use the work–energy theorem to relate the change in kinetic energy to the work done by friction. Only one force acts in the direction of motion.

EXECUTE Set the change in kinetic energy equal to the work done by friction. This results in a coefficient of kinetic friction equal to 0.50.

EVALUATE Is the change in kinetic energy and in the work negative or positive? Why?

3: Drag on an automobile

An automobile has a 150 hp engine and a top speed of 100 mph. If you assume that half of the power of the engine is delivered to the tires on the road, find the net drag (air resistance and other dissipative forces) on the automobile.

Solution Checkpoints

IDENTIFY AND SET UP At constant velocity, the net force on the car must be zero and the force acting on the tires must be equal to the drag forces. Force is related to power.

EXECUTE The power is equivalent to the force multiplied by the velocity. The units need to be converted to solve the problem:

$$100\,\text{mph} = 147\,\text{ft/s}.$$

The force is

$$F = \frac{\frac{1}{2}P}{v} = 281\,\text{lb} = 1250\,\text{N}.$$

EVALUATE Why was the power divided by 2 to get the force?

Potential Energy and Energy Conservation

Summary

In this chapter, we'll continue our investigation of energy by defining potential energy and learning about conservation of energy. Potential energy is a form of energy storage that applies to gravitational and elastic forces. Conservation of energy is one of the most fundamental concepts in physics, and we will learn how it can be applied to problems. We will learn the difference between conservative and nonconservative forces. We'll conclude by learning how to find forces, given a potential-energy function. By the end of the chapter, we'll be able to apply energy concepts to the analysis of problems and be prepared to extend our methods to additional forms of energy that we'll encounter in later chapters.

Objectives

After studying this chapter, you will understand

- The definition of potential energy.
- How to use gravitational potential energy and elastic potential energy in a variety of problems.
- The definitions of conservative and nonconservative forces.
- How to apply conservation of energy to problems.
- How to find the force, given a potential-energy function.

Concepts and Equations

Term	Description
Gravitational Potential Energy	Gravitational potential energy is the potential energy associated with the position of a particle relative to earth. For a particle of mass m at a vertical distance y above the origin in a uniform gravitational field g, the gravitational potential energy of the system is $$U_{\text{grav}} = mgy.$$ Gravitational potential energy does not depend upon the location of the origin; only *differences* in gravitational potential energy are significant.
Elastic Potential Energy	Elastic potential energy is the potential energy associated with an ideal spring. For a spring of force constant k stretched or compressed a distance x from equilibrium, the elastic potential energy is $$U_{\text{el}} = \tfrac{1}{2}kx^2.$$
Conservation of Mechanical Energy	When only conservative forces act on a particle, the total mechanical energy is constant; that is, $$K_1 + U_1 = K_2 + U_2,$$ where U is the sum of the gravitational and elastic potential energies.
Nonconservation of Mechanical Energy	When forces other than gravitation or elastic forces do work on a particle, the work W_{other} done by these other forces equals the change in the total mechanical energy: $$K_1 + U_1 + W_{\text{other}} = K_2 + U_2.$$
Conservative Forces and Conservation of Energy	Forces are either conservative or nonconservative. Conservative forces are forces for which the work–kinetic-energy theorem is completely reversible and the work can be represented by potential-energy functions. Work done by nonconservative forces manifests itself as changes in the internal energy of the object. The sum of the kinetic, potential, and internal energy is always conserved: $$\Delta K + \Delta U + \Delta U_{\text{int}} = 0.$$
Determining Force from Potential Energy	A conservative force is the negative derivative of its potential-energy function in one, two, or three dimensions: $$F_x(x) = -\frac{dU(x)}{dx},$$ $$F_x = -\frac{\partial U}{\partial x},$$ $$F_y = -\frac{\partial U}{\partial y},$$ $$F_z = -\frac{\partial U}{\partial z},$$ $$\vec{F} = -\left(\frac{\partial U}{\partial x}\hat{i} + \frac{\partial U}{\partial y}\hat{j} + \frac{\partial U}{\partial z}\hat{k}\right).$$

Conceptual Questions

1: Launching a ball

A compressed spring is used to shoot a ball straight up into the air. Compressing the spring a distance of 10 cm results in a maximum height of 3.2 m. How high does the ball go if the spring is compressed 5.0 cm?

Solution

IDENTIFY, SET UP, AND EXECUTE The spring stores elastic potential energy that is converted to gravitational potential energy at the top of the ball's flight. Elastic potential energy is proportional to the displacement squared, and gravitational potential energy is proportional to the height. One-half the compression reduces the potential energy of the spring by a factor of four, so the ball reaches one-fourth the height, or 0.8 m.

EVALUATE How does the velocity just above the spring compare for the two cases? Just above the spring, the elastic potential energy has been transformed to kinetic energy. Kinetic energy depends on the velocity squared; therefore, the velocity is proportional to the compression. One-half of the compression results in one-half the velocity just above the spring.

2: An accelerating car

A car accelerates from zero to 30 mph in 2.0 s. How long does it take to accelerate from zero to 60 mph?

Solution

IDENTIFY, SET UP, AND EXECUTE We assume that the power provided to the wheels is constant and that there is no friction. Kinetic energy is proportional to velocity squared, so a doubling of the final speed requires four times the energy. Power is energy per unit time; therefore, the time required to reach the final speed will increase by a factor of four, assuming constant power. The car will take 8.0 s to accelerate from zero to 60 mph.

EVALUATE This problem illustrates how energy principles provide alternative solutions to our earlier kinematics problems. How would you solve the problem by using forces?

3: Multiple routes to bottom of a hill

Figure 7.1 shows four different routes to the bottom of a hill, all starting from the same initial height. If you and three of your friends slide down the four routes, how do the four speeds at the bottom of the hill compare? Each of the paths is frictionless, and everyone starts from rest.

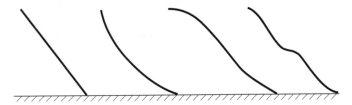

Figure 7.1 Question 3.

Solution

IDENTIFY, SET UP, AND EXECUTE Gravitational potential energy depends only on *changes* in height. Therefore, the change in gravitational potential energy is the same for all four routes, as all four have the same change in height. The kinetic energy at the bottom of the hill will be the same for all of the friends; therefore, the speeds of all four friends will be the same at the bottom.

EVALUATE If the four speeds are the same at the bottom, what quantity differs? You can see that the four routes have different lengths and are shaped differently. If you compare path 2 with path 3, you see that path 3 has a steep initial drop-off while path 2 has a shallower initial drop-off. The friend on path 3 will accelerate faster initially than the friend on path 2, will have a larger speed throughout, and will arrive at the bottom first. Therefore, the time to reach the bottom differs for the different paths.

4: Swinging by vines

Tarzan crosses a river gorge by starting from rest and swinging across the gorge on a vine. Can he ever reach a height above his starting point with this method?

Solution

IDENTIFY, SET UP, AND EXECUTE From an energy standpoint, Tarzan converts gravitational potential energy into kinetic energy as he swings across the gorge. After he passes the low point of his path, his speed slows as his gravitational potential energy increases. At his starting height, he will come to a momentary stop after all of his kinetic energy has converted to potential energy. To get to a greater final height, he needs additional energy. He could increase his initial energy by starting with an initial velocity, using a running start, for example.

EVALUATE Without additional energy, Tarzan's final height cannot be greater than his initial height.

Problems

1: Velocity of a mass on a string

A mass m is attached by a string of length l to the ceiling and is released from rest at an angle of 60° from the vertical. Find the velocity as a function of the angle.

Solution

IDENTIFY Only gravity and tension act on the mass. Tension does no work, so we can use energy conservation to solve the problem. The target variable is the velocity.

SET UP A sketch of the problem is shown in Figure 7.2. At the initial angle, the mass has only gravitational potential energy. As the mass falls, the potential energy transforms to kinetic energy. At any point, the sum of the potential and kinetic energy is equal to the initial gravitational potential energy. We will need to relate the height of the mass to the angle. The origin is placed at the bottom of the mass' path, below the anchor point.

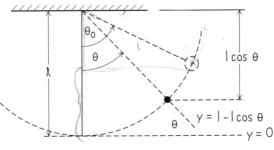

Figure 7.2 Problem 1.

EXECUTE Energy is conserved, so the initial energy is equal to the energy at any other point:

$$U_1 + K_1 = U_2 + K_2.$$

The initial kinetic energy is zero (the mass starts from rest), and the initial potential energy is

$$U_1 = mgy_1 = mg(l - l\cos\theta_0),$$

where the initial angle $\theta_0 = 60.0°$. At any other point, the kinetic and potential energies are

$$K_2 = \tfrac{1}{2}mv^2,$$
$$U_2 = mgy_2 = mg(l - l\cos\theta).$$

Equating the energies gives

$$mg(l - l\cos\theta_0) = \tfrac{1}{2}mv^2 + mg(l - l\cos\theta).$$

Rearranging to solve for v as a function of the angle gives

$$v = \sqrt{2gl(\cos\theta - \cos\theta_0)}$$
$$= \sqrt{2gl(\cos\theta - \tfrac{1}{2})}.$$

EVALUATE What is the maximum velocity? The maximum velocity occurs at the bottom of the swing, where $\theta = 0°$, and is equal to $v = \sqrt{gl}$.

2: Professor landing on spring platform

Your professor, with a mass of 60.0 kg, falls from a height of 2.50 m onto a platform mounted on a spring. As the springs compresses, she compresses the spring a maximum distance of 0.240 m. What is the force constant of the spring? Assume that the spring and platform have negligible mass.

Solution

IDENTIFY Energy is conserved, as the only forces acting on the professor are gravity and the spring force. The target variable is the force constant of the spring.

SET UP Figure 7.3 shows a sketch of the situation. Initially, the professor has only U_{grav}, since her velocity is zero ($K = 0$) and the spring is uncompressed ($U_{el} = 0$). As she falls to the top of the platform, her

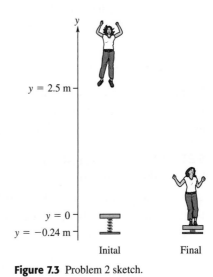

Figure 7.3 Problem 2 sketch.

kinetic energy increases and gravitational potential energy decreases. As she starts to compress the spring, she slows down as energy is transformed to the spring's elastic potential energy. At maximum compression, she comes to a momentary stop ($K = 0$). At this point, she is below the origin, so she has negative gravitational potential energy. We'll use energy conservation to solve for the spring's force constant.

EXECUTE Energy conservation relates the initial and final energies:

$$K_1 + U_1 = K_2 + U_2.$$

Initially, there is only U_{grav}. At the maximum compression, there are two potential-energy terms: U_{grav} and U_{el}. Also,

$$U_{grav,1} = U_{grav,2} + U_{el,2}.$$

Substituting the expressions for the energies yields

$$mgy_1 = mgy_2 + \tfrac{1}{2}kx^2.$$

The initial height is 2.50 m, and the final height and compression is -0.240 m. Solving for k gives

$$k = \frac{2mg(y_1 - y_2)}{x^2} = \frac{2(60.0\ \text{kg})(9.80\ \text{m/s}^2)(2.50\ \text{m} - (-0.240\ \text{m}))}{(0.240\ \text{m})^2} = 55{,}900\ \text{N/m}.$$

The force constant of the spring is 55,900 N/m.

EVALUATE Our choice of origin gave a negative y_2, but only *differences* in gravitational potential energies influence the result. This problem would have been much more challenging to solve with our force techniques, as the force of the spring varies with position.

> **CAUTION** **You set zero for gravitational potential energy!** Only *differences* in gravitational potential energy are useful in energy problems. You may set the zero at any point. It is best to choose one that simplifies the solution.

3: Designing a bungee jump

You are entering the bungee-jumping business and must design the bungee cord. The jump will be from a bridge that is 100.0 m above a river. The design calls for 2.00 seconds of free fall before the cord begins to slow the fall, and the person just touches the water after jumping. Find the force constant and length of the bungee cord for a 100.0 kg person.

Solution

IDENTIFY The forces acting on the jumper are gravity and the spring force. There is no mechanical work done on the system, so we will use energy conservation to solve the problem. Our target variables are the force constant and the length of the bungee cord.

SET UP Figure 7.4 shows a sketch of the situation with the coordinate origin at the river. On the bridge, the jumper has gravitational potential energy. After he jumps, the energy transforms to kinetic and elastic potential energies. At the river, the jumper momentarily stops and all the energy has transformed into elastic potential energy.

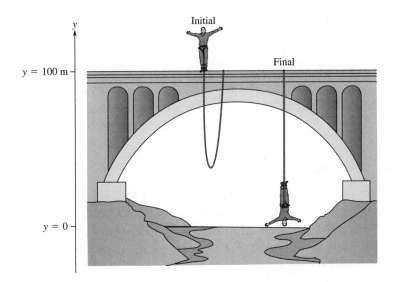

Figure 7.4 Problem 3.

We'll also need to recall our kinematics for freely falling objects to find the length of the bungee cord. The length of the bungee cord is found by determining its length when it becomes taut. We know that the cord becomes taut after 2.00 s, so we can use free-fall kinematics to solve for the distance the person falls in 2.00 s, which is equal to the length of the bungee cord. We'll ignore air resistance and any friction in the bungee cord.

EXECUTE Energy conservation relates the initial and final energies:

$$K_1 + U_1 = K_2 + U_2.$$

Initially, there is only U_{grav} at the bridge. At the river, there is only U_{el}. Hence,

$$U_{grav} = U_{el}.$$

Replacing the energies gives

$$mgy_1 = \tfrac{1}{2}kx^2,$$

where m is the mass of the jumper, y_1 is height of the bridge, x is the stretch length of the bungee cord, and k is the force constant of the bungee cord. We need the amount of stretch in the bungee cord, so we first use constant = acceleration kinematics for freely falling objects to find the position where the bungee cord becomes taut:

$$y_{taut} = y_0 + v_{0y}t + \tfrac{1}{2}a_y t^2.$$

Here, v_{0y} is zero as the person starts from rest, $y_0 = 100$ m, $a_y = -g$, and t is 2.00 s. Solving for the position where the bungee cord becomes taut, we have

$$y_{taut} = 100 \text{ m} + \tfrac{1}{2}(-9.8 \text{ m/s}^2)(2.00 \text{ s})^2 = 80.4 \text{ m}.$$

Thus, $y_{taut} = 80.4$ m is the vertical position above the river where the bungee cord becomes taut. The starting point was at $y_0 = 100$ m, so the length of the bungee cord is the difference between y_0 and y_{taut}: 100 m − 80.4 m = 19.6 m. The stretch length is how much the cord is stretched from its original length, or 80.4 m in this case (i.e., the distance from where the cord becomes taut to the river). Substituting into our energy expression to solve for the force constant yields

$$k = \frac{2mg(y_0)}{x^2} = \frac{2(100.0 \text{ kg})(9.8 \text{ m/s}^2)(100.0 \text{ m})}{(80.4 \text{ m})^2} = 30.3 \text{ N/m}.$$

You will need a bungee cord that is 19.6 m long with a 30.3 N/m force constant.

EVALUATE The spring constant was found to be relatively small, indicating that the person will be slowed gently. What will happen to a person with a mass of less than 100 kg? What about a person with a mass greater than 100 kg? The lighter person has less initial energy and so stops above the river. The heavier person has more initial energy and so stops under the surface of the river. (So the cord should be changed!)

4: Toy car loop-the-loop

A toy car is released from a spring launcher onto a horizontal track that leads to a vertical loop-the-loop, as shown in Figure 7.5. What is the minimum compression needed for the launcher so that, when released, the car remains on the track throughout the loop? The mass of the car is 10.0 g, the force constant of the launcher is 20.0 N/m, the loop has a radius of 20.0 cm, and you may assume that the car moves along the track without friction.

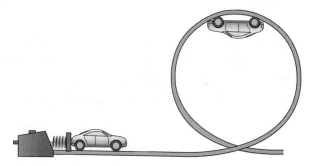

Figure 7.5 Problem 4.

Solution

IDENTIFY We will use both energy conservation and Newton's second law to solve the problem. The target variable is minimum compression of the spring.

SET UP The forces acting on the car are gravity, the spring force of the launcher while the car is in contact with the launcher, and the normal force of the ground or loop. There is no mechanical work done on the system, so energy is conserved.

For the car to remain in contact with the track at the top of the loop, it must have sufficient velocity to maintain centripetal force. A free-body diagram for the car at the top of the loop is shown in Figure 7.6.

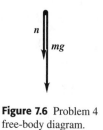

Figure 7.6 Problem 4 free-body diagram.

We place the origin at ground level. Our initial point (1) will be when the car is at rest and the launcher is compressed, storing all the energy in the spring. After the car is released, the elastic potential energy is transformed into kinetic energy and then into a combination of kinetic and gravitational potential energy when the car enters the loop. The final point (2) will be at the top of the loop, where there are both kinetic and gravitational potential energies. We'll use energy conservation to solve for the spring's compression.

EXECUTE Energy conservation relates the initial and final energies:

$$K_1 + U_1 = K_2 + U_2.$$

Initially, there is only U_{el} stored in the spring. At the top of the loop, both K and U_{grav} are stored, and

$$U_{el} = U_{grav} + K.$$

Substituting with the expressions for the energies gives

$$\tfrac{1}{2}kx^2 = mgy_2 + \tfrac{1}{2}mv^2,$$

where x is the spring compression, k is the force constant of the spring, m is the mass of the car, y_2 is twice the radius of the loop (the height at the top of the loop), and v is the speed of the car at the top of the loop. We find the velocity at the top of the loop by applying Newton's second law. To find the minimum compression of the spring, we need the minimum velocity at the top of the loop. The minimum velocity corresponds to the minimum force on the car at the top; therefore, the only force acting on the car at the top is gravity, so

$$\sum F_y = mg = ma_{rad} = \frac{mv^2}{r}.$$

Solving for v yields the velocity at the top of the loop:

$$v = \sqrt{gr} = \sqrt{(9.80 \text{ m/s}^2)(0.200 \text{ m})} = 1.40 \text{ m/s}.$$

Combining the results and solving for the displacement of the spring gives

$$x = \sqrt{\frac{m(4gr + v^2)}{k}} = \sqrt{\frac{0.0100 \text{ kg}(4(9.80 \text{ m/s}^2)(0.200 \text{ m}) + (1.40 \text{ m/s})^2)}{20.0 \text{ N/m}}} = 0.0700 \text{ m}.$$

The minimum spring compression necessary to keep the car on the track throughout the loop is 7.00 cm.

EVALUATE We found the minimum compression of the spring. Additional compression would have resulted in greater total energy after the car is launched, which would also keep the car on the track.

This problem illustrates how we'll sometimes combine our knowledge of previous materials (e.g., forces) with our current topics. As we progress through the text, we will add to our knowledge base and not merely exchange one concept for another.

5: Losing contact with the hill

A frictionless puck slides down a large, round dome of radius 2.0 m. If the puck starts at the top of the dome with a very small initial velocity, how far below the starting point does the puck lose contact with the dome?

Solution

IDENTIFY Gravitation and the normal force are the only forces acting on the puck. The normal force does no work, so we will use energy conservation. At the point where the puck loses contact with the dome, the normal force is zero. The target variable is the height at which the puck loses contact.

SET UP A sketch of the problem is shown in Figure 7.7. The origin is at the center of the dome and the target variable is the change in height, y. No mechanical work is done on the system, so energy is conserved.

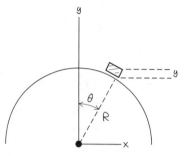

Figure 7.7 Problem 5.

The forces acting on the puck are gravity and the normal force, and the net force is a centripetal force directed toward the center of the dome. When the puck loses contact with the dome, the normal force is zero. A free-body diagram of the puck is shown in Figure 7.8.

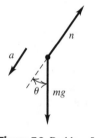

Figure 7.8 Problem 5
free-body diagram.

Our initial point (1) will be when the puck is at the top of the dome, where there is only gravitational potential energy. (We ignore the small quantity of kinetic energy at the top of the dome, to simplify the solution.) Our second point (2) will be the moment the puck leaves the dome, when there are both kinetic and gravitational potential energies.

EXECUTE We write the change in height in terms of the radius and angle:

$$y = R - R\cos\theta.$$

We need to find the angle at which the puck leaves the dome. We start with the forces acting on the puck. The net force along the radius at any point is given by Newton's second law:

$$\sum F_{rad} = mg\cos\theta - n = ma_{rad} = m\frac{v^2}{R}.$$

When the puck loses contact, the normal force goes to zero, producing the following relation between the angle and the velocity:

$$v^2 = gR\cos\theta.$$

We now have θ in terms of velocity. The v^2 reminds us of energy, so we write the expression for energy conservation:

$$U_1 + K_1 = U_2 + K_2.$$

The potential energy at the top of the dome is mgR and the kinetic energy is zero. At a later point, the puck has both kinetic energy and potential energy. Substituting in the expressions for the energies gives

$$mgR = mg(R\cos\theta) + \tfrac{1}{2}mv^2.$$

Substituting the expression from the forces results in

$$mgR = mg(R\cos\theta) + \tfrac{1}{2}mgR\cos\theta.$$

Simplifying this equation yields

$$\tfrac{2}{3}R = R\cos\theta.$$

Solving for y, we obtain

$$\begin{aligned}
y &= R - R\cos\theta \\
&= R - \tfrac{2}{3}R \\
&= \tfrac{1}{3}R = \tfrac{1}{3}(2.0\text{ m}) = 0.67\text{ m}.
\end{aligned}$$

The puck leaves the dome 0.67 m below the top.

EVALUATE We see that the result depends on neither the puck's mass nor gravity. If the puck and dome were transported to the moon, the puck would lose contact at the same position as on earth.

6: Force from potential-energy function

The potential-energy function of a particle is

$$U(x, y) = axy - 3bx^2 + 2cy^2,$$

where a, b, and c are constants. What is the force on the particle?

Solution

IDENTIFY AND SET UP Given the potential-energy function, we find the force by taking partial derivatives. The potential-energy function depends on x and y, so we will find the negative partial derivative of the potential-energy function with respect to x and y.

EXECUTE The x component of the force is

$$F_x = -\frac{\partial U}{\partial x}.$$

Substituting the expression U and solving gives

$$F_x = -\frac{\partial}{\partial x}(axy - 3bx^2 + 2cy^2) = ay - 6bx.$$

The y component of the force is

$$F_y = -\frac{\partial U}{\partial y}.$$

Substituting again and solving yields

$$F_y = -\frac{\partial}{\partial y}(axy - 3bx^2 + 2cy^2) = ax + 4cy.$$

The force is

$$\vec{F} = (ay - 6bx)\hat{i} + (ax + 4cy)\hat{j}.$$

EVALUATE This problem illustrates how we can find the force, given a potential-energy function. We see that it is easier to find the force from the potential energy than the potential energy from the force.

CAUTION **Use both forces and energy!** This problem shows how you can combine your knowledge of both forces and energy to solve complex problems. Without using the combined technique, the solution would have been more difficult.

Try It Yourself!

1: Mass on a spring

A 0.5 kg box hangs from a spring whose unstretched length is 1.0 m. The box stretches the spring 0.5 m. The box is pulled 0.5 m from its equilibrium position and is released from rest. Find its maximum velocity and height.

Solution Checkpoints

IDENTIFY AND SET UP Start by finding the spring constant by using information from the first part of the problem. What forces act on the box? Can you use energy conservation to find the position and velocity of the box at any point? The initial kinetic energy of the box is zero when it is released. What is the initial elastic potential energy?

EXECUTE Energy conservation gives

$$\tfrac{1}{2}k(1.0 \text{ m})^2 = mgy + \tfrac{1}{2}mv^2 + \tfrac{1}{2}k(1.0 \text{ m} - y)^2$$

when the gravitational potential energy is initially zero. This gives a maximum velocity of 2.2 m/s when $y = 0.5$ m and a maximum height of 1.0 m when $v = 0$.

EVALUATE Do you get the same result when you use a different origin?

2: Two masses connected by a pulley

For the frictionless system shown in Figure 7.9, find the velocity of the mass m when it hits the floor if it is released from a height L above the floor.

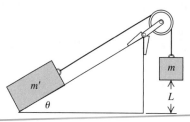

Figure 7.9 Problem 1.

Solution Checkpoints

IDENTIFY AND SET UP What forces act on the masses? Can you use energy conservation in this case? The initial kinetic energy is zero when mass m is released.

EXECUTE Energy conservation gives

$$mgL = \tfrac{1}{2}(m + m')v_2^2 + m'gL\sin\theta$$

when the origin is at the bottom of the incline. This equation results in a velocity of

$$v_2 = \sqrt{2\frac{(m - m')}{(m + m')}gL}.$$

EVALUATE What is the velocity of m as θ approaches $90°$ (an Atwood's machine)? What is the velocity as θ approaches $0°$ (a flat table)? Do these results agree with those of earlier problems in Chapter 5?

3: Force from the potential-energy function

The potential-energy function of a particle is

$$U(x) = ax + \tfrac{1}{2}kx^2.$$

What is the force on the particle? What is the equilibrium position of the particle?

Solution Checkpoints

IDENTIFY AND SET UP How do you find the force, given a potential-energy function? The equilibrium position is where the force is zero and is found by setting the force equal to zero and solving for position.

EXECUTE The force on the particle is the negative derivative with respect to position:

$$F_x = -(a + kx).$$

The force is zero when $x = -a/k$.

EVALUATE We see that this force is an elastic force.

Problem Summary

The previous two chapters have augmented our knowledge of forces and kinematics with our newly formed knowledge of energy analysis. Energy analysis shares many of the problem-solving principles we have encountered. In these problems, we

- Identified the general procedure to find the solution.
- Sketched the situation when no figure was provided.
- Identified the energies involved and the forces acting in the system.
- Applied energy principles, including conservation of energy (when possible).
- Drew free-body diagrams of the objects when appropriate.
- Applied Newton's laws when appropriate.
- Solved the equations through algebra and substitutions.
- Reflected on the results, checking for inconsistencies.

This problem-solving foundation can be applied to all problems, including those that could be solved by force analysis. Following the procedure set forth here leads to mastery of many physics problems.

8

Momentum, Impulse, and Collisions

Summary

In this chapter, we'll introduce two new concepts—*momentum* and *impulse*—which together will serve as our third major analysis technique in mechanics. Like energy analysis, momentum and impulse analysis will expand our problem-solving repertoire and allow us to tackle collision problems that would be challenging with Newton's laws. Also, like energy, momentum is a conserved quantity that has important consequences throughout physics. We'll be able to apply momentum and impulse analyses to a wide variety of problems by the end of the chapter, and, when those analyses are combined with force and energy analyses, we will be able apply a powerful set of tools to investigate many natural phenomena.

Objectives

After studying this chapter, you will understand

- The definition of momentum and the distinction between momentum and velocity.
- The definition of impulse and the distinction between impulse and force.
- How to restate Newton's law in terms of momentum and impulse.
- How to use conservation of momentum to solve a variety of collision problems.
- How to identify elastic, inelastic, and totally inelastic collisions and how to apply conservation of energy appropriately to each of these situations.
- The definition of the center of mass of a system and how to use the center of mass to solve problems.
- How to apply momentum conservation to rocket propulsion problems in which the rocket's mass changes.

Concepts and Equations

Term	Description
Momentum	The momentum $\vec{p}$ of a particle of mass m moving with velocity $\vec{v}$ is defined as $$\vec{p} = m\vec{v}.$$ Newton's second law states that the net force on a particle is equal to the rate of change of momentum of the particle: $$\sum\vec{F} = \frac{d\vec{p}}{dt}.$$ The total momentum $\vec{P}$ of a system of particles is the vector sum of the individual momenta: $$\vec{P} = \vec{p}_A + \vec{p}_B + \cdots = m_A\vec{v}_A + m_B\vec{v}_B + \cdots.$$
Impulse	The impulse $\vec{J}$ of the net force is the product of force and the time interval over which the force acts: $$\vec{J} = \vec{F}(t_2 - t_1) = \vec{F}\Delta t.$$ For net forces that vary with time, the impulse is the integral of the net force over the time interval: $$\vec{J} = \int_{t_1}^{t_2} \sum\vec{F}\,dt.$$ The change in a particle's momentum during a certain time interval equals the impulse of the net force acting on the particle during that interval: $$\vec{J} = \Delta\vec{p} = \vec{p}_2 - \vec{p}_1.$$
Conservation of Momentum	The total momentum of a system is constant when the net external force on the system is zero: $$\vec{P} = \text{constant if } \sum\vec{F} = 0.$$ Each component of momentum is separately conserved.
Elastic Collision	In an elastic collision between two bodies, the initial and final kinetic energies are equal and the initial and final relative velocities have equal magnitudes.
Inelastic Collision	In an inelastic collision between two bodies, the final kinetic energy is less than the initial kinetic energy. If the two bodies have the same final velocity, the collision is completely inelastic.
Center of Mass	The position vector $\vec{r}_{cm}$ for the center of mass of a system of particles is a weighted average of the positions $\vec{r}_1, \vec{r}_2, \ldots,$ of the individual particles: $$\vec{r}_{cm} = \frac{m_1\vec{r}_1 + m_2\vec{r}_2 + \cdots}{m_1 + m_2 + \cdots} = \frac{\sum_i m_i\vec{r}_i}{\sum_i m_i}$$
Motion of Center of Mass	The total momentum of a system equals its total mass multiplied by the velocity of the center of mass: $$\vec{P} = M\vec{v}_{cm} = m_1\vec{v}_1 + m_2\vec{v}_2 + \cdots.$$

The center of mass of a system moves as if all the mass were located at the center of mass:

$$\sum \vec{F}_{\text{ext}} = M \vec{a}_{\text{cm}}.$$

If the net external force on a system is zero, the velocity $\vec{v}_{\text{cm}}$ of the center of mass of the system is constant.

Rocket Motion	Any analysis of rocket motion must include the momentum carried away by the spent fuel and the momentum of the rocket itself.

Conceptual Questions

1: Jumping off a wall

If you fall off a 2-m-high wall, would you prefer to land on concrete or grass? Why?

Solution

Your speed at the ground will be the same in both cases. The change in your momentum, or impulse, as you come to rest will also be the same. Given the same impulse, the average force will be less if the time interval is longer. Since you will be in contact with the grass longer as you land, grass is the preferred landing material.

This conceptual question illustrates why cushioning is used to reduce the average force exerted in a collision by increasing the duration of the collision. Consider this principle when you buy your next pair of running shoes or feel your car's padded dashboard.

2: Beanbag versus tennis ball

You wish to close your bedroom door from across the room. You can toss either a beanbag or a tennis ball at the door. (Both have the same mass). Which should you choose?

Solution

Consider how much momentum each object can impart to the door. Both the beanbag and the tennis ball have the same mass, and you give each the same velocity, so their initial momenta are the same. After colliding with the door, the beanbag falls to the floor while the tennis ball bounces back toward you. The beanbag ends with zero momentum, whereas the tennis ball has momentum in the direction opposite that of its original momentum. The change in momentum of the tennis ball is larger than the change in momentum for the beanbag, so you should use the tennis ball to close the door.

Even though both objects have the same initial momentum, we see that the *change* in momentum determines the best choice.

3: Getting off the ice

You're standing on a frictionless ice rink. If you toss your physics book vertically upwards, will you move?

Solution

You will not move, since the book carries away no component of momentum parallel to the ice rink. You should toss your physics book horizontally to move along the ice.

Is momentum conserved in this case? Initially, there is no momentum. When you toss the book, you and the earth must move in the opposite direction. Since the mass of the earth is extremely large, the velocity imparted to the earth is tiny.

4: Car–train collision

If a train engine and a compact car collide, which exhibits the greatest impact force? Which exhibits the greatest change in momentum? Which exhibits the greatest acceleration?

Solution

Both the train engine and the compact car exhibit the same impact force, due to Newton's third law. Both also exhibit the same change in momentum, since the momentum is conserved. (The friction force with the ground is very small compared with the impact force and may be ignored.) However, the car exhibits the greatest acceleration, since it has the least mass.

The preferred vehicle to be in during a crash is a larger vehicle, because you will experience less acceleration (and less force will be acting on *you*). Of course, it is far better to avoid the crash altogether.

5: Collisions on a pool table

A billiard ball can stop when it collides head-on with another ball of the same mass that is at rest. Can the ball stop if the collision is at a slight angle?

Solution

If the balls collide head-on, then all of the momentum is in one direction (the direction of the initial velocity) and momentum is conserved if the second ball moves away with the first ball's initial velocity. If the balls collide at an angle, the impact gives the second ball a component of momentum perpendicular to the first ball's initial direction of motion. For the component of momentum perpendicular to the initial velocity to be conserved, the first ball must also have a perpendicular component of momentum and cannot stop.

6: Walking in a canoe

You are standing in a canoe. If you walk to the other end of the canoe, what happens to the canoe? You may neglect the resistance of the water.

Solution

No external forces act on you or the canoe (ignoring the resistance of the water). The center of mass of you and the canoe remains at rest, and its location is constant. When you walk to the other end of the canoe, the canoe must move in the opposite direction to preserve the location of the center of mass.

This conceptual question also illustrates how the frictional force is necessary for walking. As you walk, the frictional force between your shoes and the canoe pushes you in one direction while pushing the canoe in the opposite direction.

Problems

1: Tossing bubble gum

You throw your bubble gum at a stationary puck on a frictionless air-hockey table. The gum sticks to the puck, and both move away with a velocity of 1.2 m/s. If the puck has a mass of 0.30 kg and the bubble gum has a mass of 0.020 kg, find the initial speed of the bubble gum.

Solution

IDENTIFY There are no horizontal external forces, so the x component of total momentum is the same before and after the collision. The target variable is the bubble gum's initial speed.

SET UP Figure 8.1 shows the before and after sketches of the situation, including axes. Our system consists of the bubble gum and puck. Before the collision, the gum has a velocity v_{1x} and the puck is stationary. After the collision, the gum and puck move away at velocity v_{2x}. All of the velocities and momenta have only x components.

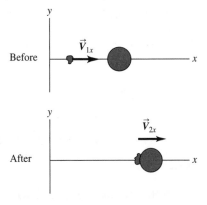

Figure 8.1 Problem 1.

EXECUTE We solve the problem by using conservation of momentum. The initial momentum is that of the gum:

$$p_{1x} = m_g v_{1x}.$$

The final momentum is that of the gum and puck together:

$$p_{2x} = (m_g + m_p)v_{2x}.$$

Momentum is conserved, so we set the momenta equal to each other:

$$m_g v_{1x} = (m_g + m_p)v_{2x}.$$

Solving for v_{1x} yields

$$v_{1x} = \frac{(m_g + m_p)v_{2x}}{m_g} = \frac{((0.020 \text{ kg}) + (0.30 \text{ kg}))(1.2 \text{ m/s})}{(0.020 \text{ kg})} = 19 \text{ m/s}.$$

The initial speed of the bubble gum is 19 m/s.

EVALUATE This problem illustrates how we can determine a projectile's velocity by examining its collision with a larger object and applying conservation of momentum.

Was the collision elastic? No, the collision was not elastic: The initial kinetic energy was 3.6 J, and the final kinetic energy was 0.23 J. This is an example of a totally inelastic collision, since the masses stuck together after the collision.

2: Ball hits the floor

A golf ball of mass 0.045 kg bounces off a tile floor. The velocity of the ball just before it hits the floor is 6.2 m/s. If the ball is in contact with the floor for 0.012 s and the floor exerts an average force of 40.0 N, find the maximum height of the ball after its impact with the floor.

Solution

IDENTIFY We can find the maximum height h by using conservation of energy, but we need to know the velocity of the ball just after impact with the floor. We are given the initial velocity and can determine the impulse imparted by the floor. Since the impulse is the change in momentum, we can find the momentum (and velocity) after the impact.

SET UP The motion is purely vertical, so we will use a single vertical axis with the positive direction taken to be upward as shown in Figure 8.2. We will first use the impulse–momentum theorem to find the velocity v_{2y} just after the ball's impact with the floor. We will then use energy conservation to find our target variable, the height h the ball reaches. The origin is located at the ground, so the golf ball only has kinetic energy immediately after the bounce. At the maximum height, the ball has pure gravitational potential energy.

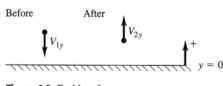

Figure 8.2 Problem 2.

EXECUTE Impulse is change in momentum, which is equal to the average force times the contact interval:

$$J_y = p_{2y} - p_{1y} = (F_{av})_y \Delta t.$$

We solve this equation for the final momentum:

$$
\begin{aligned}
p_{2y} &= (F_{av})_y \Delta t + p_{1y} \\
&= (F_{av})_y \Delta t + mv_{1y} \\
&= (40.0\ \text{N})(0.012\ \text{s}) + (0.045\ \text{kg})(-6.2\ \text{m/s}) \\
&= +0.201\ \text{kg} \cdot \text{m/s}.
\end{aligned}
$$

Note that the initial velocity is downward, or negative, with our choice of axis. The velocity just after impact is

$$v_{2y} = \frac{p_{2y}}{m} = \frac{(+0.201\ \text{kg} \cdot \text{m/s})}{(0.045\ \text{kg})} = 4.47\ \text{m/s}.$$

We can now apply conservation of energy to find the maximum height of the ball. We equate the kinetic energy of the ball just after impact with the gravitational potential energy gained by the ball when it reaches maximum height:

$$\tfrac{1}{2}mv_{2y}^2 = mgh,$$

$$h = \frac{\tfrac{1}{2}mv_{2y}^2}{mg} = \frac{v_{2y}^2}{2g} = \frac{(4.47\ \text{m/s})^2}{2(9.8\ \text{m/s}^2)} = 1.02\ \text{m}.$$

The maximum height reached by the golf ball after bouncing off of the tile floor is 1.02 m.

EVALUATE This problem illustrates the definitions of momentum and impulse in a relatively straightforward way. In addition, it reminds us that we may need to recall problem-solving skills from earlier chapters—conservation of energy in this case.

We also see how we must evaluate signs carefully, since momentum is a vector. The initial momentum is directed downward, which is negative, in our example. Had we omitted the negative sign, we would have calculated a final velocity of 16.9 m/s. Clearly, this velocity is nonsensical, since it is greater than the velocity just before the bounce!

3: Stepping off a sled

A 60.0 kg sled is traveling across an ice rink at 4.5 m/s. Irene, riding on the sled, jumps off and lands on the ice with a velocity of 1.5 m/s in the opposite direction. What is the velocity of the sled after Irene jumps? Irene has a mass of 45.0 kg.

Solution

IDENTIFY There are no external horizontal forces acting on the sled-plus-Irene system, so momentum is conserved.

SET UP We take the x-axis to lie along the direction of motion of the sled, with the positive direction taken to be in the initial direction of the sled as shown in Figure 8.3. We are given the masses of the sled and Irene, the initial velocity of the sled (with Irene on it), and Irene's final velocity. Our target variable is v_{S2x}, the final velocity of the sled.

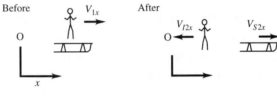

Figure 8.3 Problem 3.

EXECUTE The x-component of total momentum before Irene jumps off is

$$p_{1x} = m_S v_{S1x} + m_I v_{I1x}$$
$$= (60.0 \text{ kg})(4.5 \text{ m/s}) + (45.0 \text{ kg})(4.5 \text{ m/s})$$
$$= 472.5 \text{ kg} \cdot \text{m/s}.$$

(Note again that both the sled and Irene are moving with the same initial velocity.) After Irene jumps off, the x component of total momentum has the same value, so

$$p_{2x} = m_S v_{S2x} + m_I v_{I2x}.$$

Solving for v_{S2x}, gives

$$v_{S2x} = \frac{p_{1x} - m_I v_{I2x}}{m_S}$$
$$= \frac{(472.5 \text{ kg} \cdot \text{m/s}) - (45.0 \text{ kg})(-1.5 \text{ m/s})}{(60.0 \text{ kg})}$$
$$= 9.0 \text{ m/s},$$

where Irene's final velocity is negative, since she ends up moving away from the sled. The final velocity of the sled is 9.0 m/s in the positive direction.

EVALUATE We check the result by noting that the final velocity of the sled is greater than the initial velocity. For Irene to move away from the sled, she had to give the cart an impulse that resulted in a larger final velocity.

This problem illustrates how we can apply conservation of momentum to problems involving no net external force. Here, there is no explicit collision, but simply someone jumping off a sled.

CAUTION **Watch out for signs!** In the last two problems, we paid careful attention to the direction, and therefore the sign, of the momentum. Note how an incorrect sign could have led to nonsensical results. Momentum is a vector, and you must always check and recheck the direction to ensure accurate results.

4: Colliding pucks in two dimensions

Two pucks collide on a frictionless air-hockey table. Initially, puck A is traveling at 3.50 m/s and puck B is at rest. After the collision, puck A moves away at a speed of 2.50 m/s and an angle of 30.0° from the initial direction. Find the final velocity of puck B. Puck A has a mass of 3.0 kg and puck B has a mass of 5.0 kg.

Solution

IDENTIFY There are no horizontal external forces, so both the x component and the y component of the total momentum are conserved in the collision.

SET UP Figure 8.4 shows the before and after sketches of the situation, including axes. Our system consists of the two pucks. Before the collision, puck A moves with velocity v_{A1} to the right. After the collision, puck A moves away at velocity v_{A2} 30.0° above the x-axis and puck B moves away at velocity v_{B2} at an angle θ below the x-axis (to conserve momentum). The velocities are not along a single axis, so we will have to solve for both the x and y components of momentum. Our target variable is the final velocity v_{B2} of puck B.

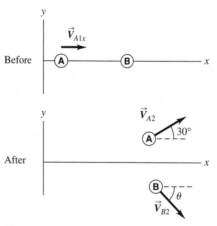

Figure 8.4 Problem 4.

EXECUTE Starting with the x components of momentum before and after the collision, we have

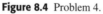

$$p_{1x} = m_A v_{A1x},$$
$$p_{2x} = m_A v_{A2x} + m_B v_{B2x} = m_A v_{A2} \cos 30.0° + m_B v_{B2x}.$$

The x component of momentum is conserved, so we set the expressions for p_{1x} and p_{2x} equal to each other:

$$m_A v_{A1x} = m_A v_{A2} \cos 30.0° + m_B v_{B2x}.$$

Solving for v_{B2x} yields

$$v_{B2x} = \frac{m_A v_{A1x} - m_A v_{A2} \cos 30.0°}{m_B}$$

$$= \frac{(3.0\,\text{kg})(3.50\,\text{m/s}) - (3.0\,\text{kg})(2.50\,\text{m/s})\cos 30.0°}{(5.0\,\text{kg})}$$

$$= 0.801\,\text{m/s}.$$

We follow the same procedure for the y components of momentum. The initial y component of momentum is zero. The y components of the final momentum must sum to zero; that is,

$$0 = p_{2y} = m_A v_{A2y} + m_B v_{B2y} = m_A v_{A2} \sin 30.0° + m_B v_{B2y}.$$

Solving for v_{B2y} gives

$$v_{B2y} = \frac{-m_A v_{A2} \sin 30.0°}{m_B}$$

$$= \frac{-(2.50\,\text{m/s})(3.0\,\text{kg})\sin 30.0°}{(5.0\,\text{kg})}$$

$$= -0.75\,\text{m/s}.$$

The x and y components of puck B's velocity are 0.80 m/s and -0.75 m/s, respectively. Thus, puck B must travel into the fourth quadrant, as expected. We find the magnitude and direction of puck B's velocity from

$$v_{B2} = \sqrt{v_{B2x}^2 + v_{B2y}^2} = \sqrt{(0.80\,\text{m/s})^2 + (-0.75\,\text{m/s})^2} = 1.10\,\text{m/s},$$

$$\phi = \tan^{-1}\frac{v_{B2y}}{v_{B2x}} = \tan^{-1}\frac{(-0.75\,\text{m/s})^2}{(0.80\,\text{m/s})} = -43.2°,$$

Puck B's final velocity is 1.10 m/s, directed at an angle of 43.2° below the positive x-axis.

EVALUATE We can check our answer by examining the momentum components before and after the collision. Initially, puck A has an x component of momentum equal to 10.5 kg m/s. After the collision, the x component of momentum of puck A is 6.5 kg m/s and that of puck B is 4.0 kg m/s, or a total of 10.5 kg m/s, as expected. There is no initial momentum along the y-axis. After the collision, the y component of momentum of puck A is 3.75 kg m/s and that of puck B is -3.75 kg m/s, also as expected.

Solving momentum problems in two dimensions follows from the one-dimensional cases. You just need to remember that momentum is a vector that can have multiple components, each of which can be conserved individually.

5: Elastic collision in one dimension

Two gliders collide elastically on a frictionless, linear air track. Glider A has a mass of 0.60 kg and initially moves to the right at 3.0 m/s. Glider B has mass of 0.40 kg and initially moves to the left at 4.0 m/s. What are the final velocities of the two gliders after the collision?

Solution

IDENTIFY There are no net external forces acting on the system, so the momentum of the system is conserved. The collision is elastic; therefore, energy is conserved as well.

SET UP Figure 8.5 shows the before and after sketches of the situation including axes. Our system consists of the two gliders. Before the collision, glider A moves with velocity $v_{A1x} = 3.0$ m/s to the right and glider B moves with velocity $v_{B1x} = -4.0$ m/s to the left. After the collision, glider A moves away with velocity v_{A2x} and glider B moves away with velocity v_{B2x}. Our target variables are the final velocities of the two gliders. We'll need to use the relative velocity relation for elastic collisions in our solution.

Before

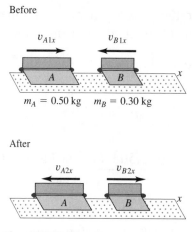

v_{A1x} v_{B1x}

$m_A = 0.50$ kg $m_B = 0.30$ kg

After

v_{A2x} v_{B2x}

Figure 8.5 Problem 5.

EXECUTE From conservation of momentum,

$$m_A v_{A1x} + m_B v_{B1x} = m_A v_{A2x} + m_B v_{B2x},$$
$$(0.60\text{ kg})(3.0\text{ m/s}) + (0.40\text{ kg})(-4.0\text{ m/s}) = (0.60\text{ kg})v_{A2x} + (0.40\text{ kg})v_{B2x},$$
$$0.20\text{ m/s} = 0.60v_{A2x} + 0.40v_{B2x}.$$

The last equation has two unknowns, and we need more information to solve for the velocities. Since this is an elastic collision, we can apply the relative velocity relation to solve for the velocities:

$$v_{B2x} - v_{A2x} = -(v_{B1x} - v_{A1x})$$
$$v_{B2x} - v_{A2x} = -((-4.0\text{ m/s}) - (3.0\text{ m/s})) = 7.0\text{ m/s}$$
$$v_{B2x} = v_{A2x} + 7.0\text{ m/s}.$$

Substituting the last expression into the earlier momentum conservation relation gives

$$0.60v_{A2x} + 0.40(v_{A2x} + 7.0\text{ m/s}) = 0.20\text{ m/s}$$
$$v_{A2x} = (0.20\text{ m/s}) - (0.40(7.0\text{ m/s})) = -2.6\text{ m/s}$$
$$v_{B2x} = v_{A2x} + 7.0\text{ m/s} = (-2.6\text{ m/s}) + 7.0\text{ m/s} = 4.4\text{ m/s}.$$

After the collision, puck A moves to the left at 2.6 m/s (v_{A2x} is negative) and puck B moves to the right at 4.4 m/s.

EVALUATE Both gliders reversed their directions in this elastic collision. Are the kinetic energies equivalent before and after the collision? The initial kinetic energy is

$$K_i = \tfrac{1}{2}m_A v_{A1x}^2 + \tfrac{1}{2}m_B v_{B1x}^2 = \tfrac{1}{2}(0.60\text{ kg})(3.0\text{ m/s})^2 + \tfrac{1}{2}(0.40\text{ kg})(-4.0\text{ m/s})^2 = 5.9\text{ J}.$$

The final kinetic energy is

$$K_f = \tfrac{1}{2}m_A v_{A2x}^2 + \tfrac{1}{2}m_B v_{B2x}^2 = \tfrac{1}{2}(0.60\ \text{kg})(-2.6\ \text{m/s})^2 + \tfrac{1}{2}(0.40\ \text{kg})(4.4\ \text{m/s})^2 = 5.9\ \text{J}.$$

The initial and final kinetic energies are equivalent, as expected.

6: Ballistic pendulum

A 6.00 g bullet is shot through a 2.00 kg block suspended on a 1.00-m-long string. If the initial speed of the bullet is 650 m/s and it emerges with a velocity of 175 m/s, find the maximum angle through which the block swings after it is hit.

Solution

IDENTIFY We will analyze this problem in two stages. First, we'll look at the interaction of the bullet with the block. Second, we'll examine the swinging of the block on the string after the bullet passes through the block.

In the first stage, the bullet passes through the block, giving the block an impulse. The bullet passes through the block very quickly, so the block has no time to swing any significant distance from its initial position. The only force acting on the bullet-and-block system is between the bullet and block; there are no appreciable external forces acting on the system. We conclude that, during the first stage, the horizontal component of momentum is conserved.

In the second stage, the block moves away from its initial position. The only forces acting on the block are gravity (a conserved force) and tensions due to the strings (which do no work as the block swings). Mechanical energy is conserved as the block swings.

SET UP Figure 8.6 shows sketches of the two stages of the problem. We take the positive x-axis to be to the right and the positive y-axis upward. Our target variable for the first stage is the velocity V_{2x} of the block after the bullet emerges from the block. We will use momentum conservation to find V_{2x}. Our target variable for the second stage is θ, the angle the string makes with the vertical when the block stops momentarily after swinging to the right. We will use energy conservation to find the height h the block rises to after the collision and relate h to θ.

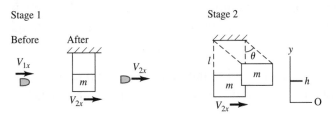

Figure 8.6 Problem 6.

EXECUTE In the first stage, conservation of momentum gives

$$m v_{1x} = m v_{2x} + M V_{2x}.$$

The velocity of the block is

$$
\begin{aligned}
V_{2x} &= \frac{m(v_{1x} - v_{2x})}{M} \\
&= \frac{(0.006\ \text{kg})(650\ \text{m/s} - 175\ \text{m/s})}{(2.0\ \text{kg})} \\
&= 1.43\ \text{m/s}.
\end{aligned}
$$

At the beginning of the second stage, the block has kinetic energy. Afterwards, the block swings up and comes to rest momentarily. At this point, the block's kinetic energy is zero and its gravitational potential energy has increased by mgh. Energy conservation gives

$$\tfrac{1}{2}MV_{2x}^2 = Mgh.$$

The block reaches a height

$$h = \frac{V_{2x}^2}{2g} = 0.104 \text{ m.}$$

The angle between the vertical and the string when the block stops momentarily is

$$\theta = \cos^{-1}\frac{l - h}{l} = 26.4°.$$

The maximum angle the block swings to after the collision is 26.4°.

EVALUATE Where did most of the energy go? A quick check of the kinetic energies gives an initial energy of the bullet (before striking the block) of 1270 J, a final energy of the bullet of 92 J, and a final energy of the block of 2.0 J. Most of the initial energy was dissipated in the deformation and heating of the bullet and block.

7: Boat on a lake

Luka is standing in a boat on a calm lake. His position is 5.00 m from the end of the pier. As he walks towards the pier, the boat moves away from it. When Luka stops walking, he finds that the boat has moved 2.00 m away from the pier. If you ignore the boat's resistance to motion in the water, how far from the end of the pier is Luka when he stops walking? The mass of the boat is 80.0 kg, and Luka's mass is 60.0 kg.

Solution

IDENTIFY Since we can ignore friction between the boat and the water, the net external force on the boat-and-Luka system is zero. Momentum is therefore conserved. There is no initial motion, so the total momentum is zero and the velocity of the center of mass of the system is zero. The center of mass will remain at rest as Luka walks in the boat.

SET UP We take the origin to be the end of the pier, since the measurements are given with respect to that point. Figure 8.7 shows a sketch of the situation. Our target variable is the final position of Luka, x_{L2}.

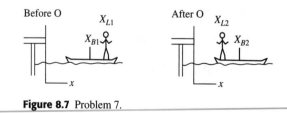

Figure 8.7 Problem 7.

EXECUTE The initial coordinate of the center of mass of the boat-and-Luka system is given by the formula

$$x_{cm} = \frac{x_{L1}m_L + x_{B1}m_B}{m_L + m_B}.$$

The final coordinate of the center of mass of the system is given by the equation

$$x_{cm} = \frac{x_{L2}m_L + x_{B2}m_B}{m_L + m_B}.$$

Since the center of mass doesn't move, we set the two right-hand expressions equal to each other:

$$\frac{x_{L1}m_L + x_{B1}m_B}{m_L + m_B} = \frac{x_{L2}m_L + x_{B2}m_B}{m_L + m_B}.$$

The boat moves 2.0 m away from the end of the pier, so $x_{B2} = x_{B1} + 2.0$ m. Substituting and solving gives

$$x_{L1}m_L + x_{B1}m_B = x_{L2}m_L + (x_{B1} + 2.0\,\text{m})m_B,$$
$$(5.0\,\text{m})(60.0\,\text{kg}) + x_{B1}(80.0\,\text{kg}) = x_{L2}(60.0\,\text{kg}) + (x_{B1} + 2.0\,\text{m})(80.0\,\text{kg}),$$
$$x_{L2} = 2.3\,\text{m}.$$

EVALUATE Since the boat and Luka's mass are similar, Luka moves a distance similar to the distance the boat moves. Luka's mass is less, so he moves a little farther than the boat moves.

8: Falling sand

Sand is dropped from a height of 1.0 m onto a kitchen scale at the uniform rate of 100.0 g/s. Find the force on the scale and its reading if the scale is calibrated in kg.

Solution

IDENTIFY We need to find the force on the scale due to the falling sand. We can find the change in momentum of the falling sand and relate it to the force through impulse. The target variable is the force on the scale.

SET UP We will first find the velocity of the sand as it hits the scale. We will then use that velocity to find the change in momentum as the sand strikes the scale. We will incorporate the change in momentum into the impulse formula and solve for the force on the scale. We will complete the problem by converting the force into a calibrated scale reading. We will take our vertical axis to be positive downward.

EXECUTE As the sand falls 1.0 m, it acquires a velocity

$$v_{1y} = \sqrt{2gh},$$

according to energy conservation. The sand stops after striking the scale. The change in momentum as the sand strikes the scale is

$$\Delta p = p_{2y} - p_{1y} = -\Delta mv = -\Delta m\sqrt{2gh}.$$

The change in momentum is negative (directed upward) in our coordinate system. The force on the sand due to the scale is directed upward to stop the sand. The impulse imparted to the sand by the scale is

$$J = -F\Delta t = \Delta p = -\Delta m\sqrt{2gh}.$$

Rearranging terms and dividing both sides by Δt gives

$$F = \frac{\Delta m}{\Delta t}\sqrt{2gh}$$
$$= (0.100 \text{ kg/s})\sqrt{2(9.8 \text{ m/s}^2)(1.0 \text{ m})}$$
$$= 0.44 \text{ N}.$$

Here, the rate of change of mass per unit time is the given rate at which the sand falls. The scale is calibrated in terms of mass, so we divide the force by g to get the reading on the scale:

$$m = \frac{F}{g} = \frac{0.44 \text{ N}}{9.8 \text{ m/s}^2} = 0.045 \text{ kg}.$$

The scale reads 0.045 kg as the sand falls.

EVALUATE This problem illustrates how we can use force, momentum, and impulse together to solve problems.

Try It Yourself!

1: Rocket motion

A two-stage rocket traveling at 350 m/s through space separates, with one stage having twice the mass of the other. If the final velocity of the larger stage is 120 m/s in a direction opposite its initial direction, find the final velocity of the smaller stage.

Solution Checkpoints

IDENTIFY AND SET UP Confirm that there are no net external forces acting on the rocket. Apply conservation of momentum along one axis. The target variable is the velocity of the smaller rocket stage.

EXECUTE Momentum conservation gives

$$m_A v_{1x} + m_B v_{1x} = m_A v_{A2x} + m_B v_{B2x}.$$

Solving for v_{A2x} yields

$$v_{A2x} = \frac{(m_A + m_B)v_{1x} - m_B v_{B2x}}{m_A} = 1300 \text{ m/s}.$$

EVALUATE Do you expect the smaller stage to have a final velocity that is larger or smaller than the other velocities in the problem? Do you expect energy to be conserved in this case?

2: Car collision

A 2000 kg car moving at 30 km/hr collides with a stopped 1000 kg car, and the two lock bumpers. What is their common velocity after the collision? What fraction of the initial energy is dissipated in the collision?

Solution Checkpoints

IDENTIFY AND SET UP Confirm that there are no net external forces acting on the cars just before and just after the collision. Apply conservation of momentum along one axis. The target variables are the velocity of the combined cars and the fraction of energy lost in the collision.

EXECUTE Momentum conservation gives

$$m_A v_{A1x} = (m_A + m_B)v_{2x}.$$

Solving for the final velocity gives 20 km/hr. The initial and final energies are

$$E_1 = \tfrac{1}{2}m_A v_{A1x}^2$$
$$E_2 = \tfrac{1}{2}(m_A + m_B)v_{2x}^2.$$

Dividing the two demonstrates that two-thirds of the initial energy remains kinetic after the collision.

EVALUATE We will contrast this problem with the next one, in which the collision is elastic.

3: Car collision revisited

A 2000 kg car moving at 30 km/hr collides with a stopped 1000 kg car, and the two have spring bumpers so that the collision is perfectly elastic. What is the velocity of each car after the collision?

Solution Checkpoints

IDENTIFY AND SET UP Confirm that there are no net external forces acting on the cars just before and just after the collision. Apply conservation of momentum along one axis. Apply conservation of energy, since the collision is elastic. The target variables are the velocities of the two cars.

EXECUTE Momentum conservation gives

$$m_A v_{A1x} = m_A v_{A2x} + m_B v_{B2x}.$$

Energy conservation gives

$$\tfrac{1}{2}m_A v_{A1x}^2 = \tfrac{1}{2}m_A v_{A2x}^2 + \tfrac{1}{2}m_B v_{B2x}^2.$$

These two equations can be solved to find $v_{A2x} = 10$ km/hr and $v_{2Bx} = 40$ km/hr. You may wish to consult the text for a derivation of velocities in elastic collisions.

EVALUATE In this elastic collision, we see that the lighter car ends up with a velocity greater than the heavier car after the collision. You can also check that the initial and final momenta and energies are equal.

4: Putty hitting mass on spring

A 0.75 kg mass of putty is hurled with a velocity of 2.0 m/s against a 0.50 kg mass attached to a spring, as shown in Figure 8.8. The mass attached to the spring slides across the frictionless horizontal surface, depressing the spring a maximum distance of 12.0 cm. Find the spring constant if the putty sticks to the mass on the spring.

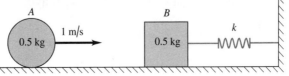

Figure 8.8 Try It Yourself! 4.

Solution Checkpoints

IDENTIFY AND SET UP Is the collision elastic or inelastic if the putty sticks to the mass? You can apply conservation of momentum to the initial collision (before the spring begins to compress). Then apply conservation of energy to the interval after the spring begins to compress.

EXECUTE Momentum conservation gives

$$m_A v_{A1x} = (m_A + m_B)v_{2x}.$$

The velocity of the putty plus mass is 1.2 m/s just after the collision. Energy conservation applied to the interval after the collision gives

$$\tfrac{1}{2}(m_A + m_B)v_{2x}^2 = \tfrac{1}{2}kx^2.$$

Substituting yields a spring constant of 125 N/m.

EVALUATE Energy and momentum conservation do not apply to the whole process; rather, the process must be broken into intervals during which energy and momentum can be applied. Can you repeat the problem, replacing the putty for a rubber ball that bounces elastically off the mass attached to the spring?

5: Collision on a football field

A 135 kg football player traveling at 5.0 m/s collides with an 85 kg player at rest. The two slide 2.0 m on wet grass. Their collision is completely inelastic. What is the coefficient of friction between the grass and the players? How much energy is dissipated in the collision?

Solution Checkpoints

IDENTIFY AND SET UP Confirm that there are minimal external forces acting on the players just before and just after the collision. Apply conservation of momentum to the two players as they collide. Then use the work–energy theorem to find the friction force. The energy dissipated is found by comparing the energy just before and just after the collision.

EXECUTE Momentum conservation applied to the collision gives

$$m_A v_{A1x} = (m_A + m_B)v_{2x}.$$

The work–energy theorem applied to the two players as they slide on the grass after they collide results in

$$-\mu(m_A + m_B)gx = 0 - \tfrac{1}{2}(m_A + m_B)v_{2x}^2.$$

Solving these equations, we obtain $\mu = 0.24$. The change in energy during the collision is

$$\Delta E = \tfrac{1}{2}(m_A + m_B)v_{2x}^2 - \tfrac{1}{2}m_A v_{A1x}^2,$$

or 650 J dissipated during the collision.

EVALUATE In this collision, we had to split the process up into two intervals, to which we separately applied conservation of momentum and the work–energy theorem. We also found that almost one-third of the energy is dissipated in the totally inelastic collision of the players.

Problem Summary

This chapter has augmented our knowledge of energy, forces, and kinematics with our newly formed knowledge of momentum analysis. Momentum analysis shares many of the problem-solving principles we have encountered. In the problems presented, we

- Identified the general procedure to find the solution.
- Sketched the situation, including before and after views.
- Identified the momenta, energies, and forces in the system.
- Applied conservation of momentum to the system.
- Applied energy principles, including conservation of energy (when possible).
- Drew free-body diagrams of the objects when appropriate.
- Applied Newton's laws when appropriate.
- Solved for the target variable(s) algebraically from the equations we derived.
- Reflected on the results, checking for inconsistencies.

9 Rotation of Rigid Bodies

Summary

In this chapter, we will investigate the rotational motion of *rigid bodies*—objects that don't change size or shape as they move. We'll describe first the kinematics of rotation for a rigid body and then its rotational kinetic energy. We'll see how these quantities are analogous to linear kinematics and translational kinetic energy. Finally, we'll learn about the moment of inertia, how to calculate it, and how to use it to measure rotational inertia. We'll use our new knowledge of rotational motion in the next chapter as we learn how to cause such motion.

Objectives

After studying this chapter, you will understand

- How to use radians in angular measurements.
- The definition and application of angular displacement, velocity, and acceleration.
- How to solve problems involving constant angular acceleration.
- How to define moment of inertia and apply it to systems of varying shapes.
- How to calculate the moment of inertia.
- How to solve conservation-of-energy problems that include rotational kinetic energy.
- How to draw analogies between translational and rotational motion and energy.

Concepts and Equations

Term	Description
Rigid Body	A rigid body is an object that maintains an unchanging size and shape. We neglect squeezing, stretching, and twisting in our analysis of a rigid body.
Radian	Angular displacements are usually measured in radians. A displacement θ measured in radians is the ratio of the arc length s to the radius r: $$\theta = \frac{s}{r}.$$ There are 2π radians in one revolution $(360°)$.
Angular Velocity	The instantaneous angular velocity about the z-axis is the rate of change of angular displacement with respect to time: $$\omega_z = \lim_{\Delta t \to 0} \frac{\Delta \theta}{\Delta t} = \frac{d\theta}{dt}.$$ The term *angular velocity* refers to the instantaneous angular velocity. All pieces of a rigid object have the same angular velocity at any given instant. The direction of the angular velocity is given by the right-hand rule.
Angular Acceleration	The instantaneous angular acceleration about the z-axis is the rate of change of angular velocity with respect to time: $$\alpha_z = \lim_{\Delta t \to 0} \frac{\Delta \omega_z}{\Delta t} = \frac{d\omega_z}{dt} = \frac{d^2\theta}{dt^2}.$$ The term *angular acceleration* refers to the instantaneous angular acceleration. All pieces of a rigid object have the same angular acceleration at any given instant. The direction of the angular acceleration is the same as that of the angular speed when the object is speeding up and opposite that of the angular speed when the object is slowing down.
Rotation with Constant Angular Acceleration	When an object moves with constant angular acceleration, the angular displacement, velocity, acceleration, and time are related by the formulas $$\theta = \theta_0 + \omega_{0z}t + \tfrac{1}{2}\alpha_z t^2,$$ $$\omega_z = \omega_{0z} + \alpha_z t,$$ $$\omega_z^2 = \omega_{0z}^2 + 2\alpha_z(\theta - \theta_0),$$ $$\theta - \theta_0 = \tfrac{1}{2}(\omega_z + \omega_{0z})t,$$ where θ_0 and ω_{0x} are the initial values of the angular position and velocity, respectively.
Connecting Linear and Angular Quantities	The tangential speed v of a particle rotating in a rigid body at a distance r from the axis of rotation is $$v = r\omega.$$ The particle's acceleration $\vec{a}$ has a tangential component $$a_{\text{tan}} = \frac{dv}{dt} = \frac{rd\omega}{dt} = r\alpha$$ and a radial component $$a_{\text{rad}} = \frac{v^2}{r} = \omega^2 r.$$

Moment of Inertia	The moment of inertia, I, of a body is a measure of its rotational inertia and depends on how the body's mass is distributed relative to the axis of rotation. The moment of inertia is given by $$I = m_1 r_1^2 + m_2 r_2^2 + m_3 r_3^2 + \cdots$$ $$= \sum_i m_i r_i^2.$$ For arbitrarily distributed masses, the moment of inertia is given by $$I = \int r^2 dm.$$ Moments of inertia for common shapes are given in Table 9.2 in the textbook. The parallel-axis theorem can be used to find the moment of inertia, I_P, about another parallel axis; that is, $$I_P = I_{\text{cm}} + Md^2,$$ where I_{cm} is the moment of inertia about the center of mass, M is the mass of the body, and d is the distance between the two axes.
Energy of Rotating Body	The kinetic energy of a rigid body rotating about a fixed axis is $$K = \tfrac{1}{2} I \omega^2.$$ This quantity is the sum of the kinetic energies of all of the particles that make up the rigid body.

Conceptual Questions

1: Rolling versus sliding down a hill

A ball travels down a hill. Will the ball reach the bottom of the hill faster if it rolls or if it slides without friction down the hill?

Solution

IDENTIFY, SET UP, AND EXECUTE No energy is lost as the ball travels down the hill, so we can use energy conservation to answer this question. At the top of the hill, the ball has gravitational potential energy. As the ball descends, the gravitational potential energy is transformed into kinetic energy. When the ball rolls down the hill, the kinetic energy is shared between translational kinetic energy and rotational kinetic energy. When the ball slides down the hill without friction, the gravitational potential energy transforms into translational kinetic energy alone. There is no rotational kinetic energy in this case. Therefore, more energy is transformed into translational kinetic energy if the ball slides without friction than if it rolls down the hill. Since it acquires more translational kinetic energy, its velocity is higher and it reaches the bottom faster when it slides without friction.

EVALUATE This question shows how we must include both translational and rotational kinetic energy in our energy analyses. We'll practice using rotational kinetic energy in the problem section.

2: Comparing moments of inertia

A light rod of length L has two lead weights of mass M attached at both ends of the rod. How does the system's moment of inertia compare when the rod is spun about an axis at its center as opposed to when it is spun around a point one-quarter along the length of the rod?

IDENTIFY, SET UP, AND EXECUTE The moment of inertia of an object is

$$I = m_1 r_1^2 + m_2 r_2^2 + \cdots.$$

The moment of inertia when the axis is at the center of the rod is then

$$I_{\text{center axis}} = M\left(\frac{L}{2}\right)^2 + M\left(\frac{L}{2}\right)^2 = M\left(\frac{L^2}{2}\right) = \tfrac{1}{2}ML^2,$$

since each mass is positioned half of the length of the rod away from the axis. When the axis is one-quarter along the length of the rod, the moment of inertia is

$$I_{\text{1/4 along rod}} = M\left(\frac{L}{4}\right)^2 + M\left(\frac{3L}{4}\right)^2 = M\left(\frac{L^2}{16}\right) + M\left(\frac{9L^2}{16}\right) = \tfrac{10}{16}ML^2,$$

since one mass is $\frac{1}{4}L$ from the axis and the other is $\frac{3}{4}L$ from the axis. The moment of inertia when the rod is spun one-quarter along its length is 25% larger than the moment of inertia when the rod is spun at the center.

EVALUATE The moment of inertia depends on both mass and the location of the mass. Since the location from the axis enters as the square of the distance, moving the axis changes the moment of inertia. Do you get the same result if you apply the parallel-axis theorem?

Practice Problem: What axis of rotation provides the largest moment of inertia? *Answer*: The axis located at one end of the rod gives the largest moment of inertia, $I = ML^2$.

3: Rolling up a ramp

A solid sphere and a thin-walled sphere roll without slipping along a horizontal surface. The two spheres roll with the same translational speed. The surface leads to a ramp. Which sphere rises to the greatest height on the ramp before stopping momentarily?

IDENTIFY AND SET UP We'll use conservation of energy and ignore air drag. Both spheres have initial translational and rotational kinetic energies that are transformed completely into gravitational potential energy when they stop momentarily on the ramp. The sphere with the greatest initial total kinetic energy will rise to the greatest height.

EXECUTE The initial kinetic energy of each sphere is

$$K_i = \tfrac{1}{2}m_{\text{sphere}}v_{\text{cm}}^2 + \tfrac{1}{2}I_{\text{sphere}}\omega^2.$$

The angular velocity is related to the velocity of the center of mass, since both spheres roll without slipping. Thus,

$$\omega = \frac{v_{\text{cm}}}{R},$$

where R is the radius of the sphere. The moment of inertia of the solid sphere is $I_{\text{solid}} = 2/5 m_{\text{solid}} R_{\text{solid}}^2$. The kinetic energy of the solid sphere is then

$$K_{\text{solid}} = \tfrac{1}{2}m_{\text{solid}}v_{\text{cm}}^2 + \tfrac{1}{2}I_{\text{solid}}\omega^2 = \tfrac{1}{2}m_{\text{solid}}v_{\text{cm}}^2 + \tfrac{1}{2}\tfrac{2}{5}m_{\text{solid}}R_{\text{solid}}^2\left(\frac{v_{\text{cm}}}{R_{\text{solid}}}\right)^2 = \tfrac{7}{10}m_{\text{solid}}v_{\text{cm}}^2.$$

The moment of inertia of the thin-walled sphere (shell) is $I_{shell} = 2/3 m_{shell} R_{shell}^2$. The kinetic energy of the shell is then

$$K_{shell} = \tfrac{1}{2} m_{shell} v_{cm}^2 + \tfrac{1}{2} I_{shell} \omega^2 = \tfrac{1}{2} m_{shell} v_{cm}^2 + \tfrac{1}{2} \tfrac{2}{3} m_{shell} R_{shell}^2 \left(\frac{v_{cm}}{R_{shell}} \right)^2 = \tfrac{5}{6} m_{shell} v_{cm}^2.$$

Since the final gravitational potential energy depends on mass ($U = mgh$), the masses will cancel and we compare the leading fractions in the kinetic-energy terms to determine which sphere has the greatest initial kinetic energy. We see that the shell has more initial kinetic energy; therefore, the thin-walled sphere rises to the greatest height.

EVALUATE It is interesting to find that the results don't depend on either the mass or the radius of the two spheres. The results depend only on how the mass is distributed in the object.

4: Racing down a ramp

A thin-walled hollow cylinder, a solid cylinder, a solid sphere, and a thin-walled sphere start from rest at the same height and roll without slipping down a wide ramp. Rank the velocities of the four objects, from first to last.

IDENTIFY AND SET UP We'll use conservation of energy. All of the objects have gravitational potential energy that is transformed into translational and rotational kinetic energies. We will find the velocity at the bottom of the ramp in terms of the height and other factors.

EXECUTE For any of the four objects, energy conservation applied to the starting point and the bottom of the ramp gives

$$mgh = \tfrac{1}{2} m v^2 + \tfrac{1}{2} I \omega^2.$$

The angular velocity is related to the velocity of the center of mass, since all of the objects roll without slipping; thus,

$$\omega = \frac{v}{R},$$

where R is the radius of the object. Replacing the angular velocity yields

$$mgh = \tfrac{1}{2} m v^2 + \tfrac{1}{2} I \left(\frac{v}{R} \right)^2.$$

We now solve for the velocity of each of the four objects. The thin-walled hollow cylinder (TWHC) gives

$$m_{TWHC} gh = \tfrac{1}{2} m_{TWHC} v_{TWHC}^2 + \tfrac{1}{2} \left(m_{TWHC} R^2 \right) \left(\frac{v_{TWHC}}{R} \right)^2 = m_{TWHC} v_{TWHC}^2,$$

$$v_{TWHC} = \sqrt{gh}.$$

The solid cylinder (SC) results in

$$m_{SC} gh = \tfrac{1}{2} m_{SC} v_{SC}^2 + \tfrac{1}{2} \left(\tfrac{1}{2} m_{SC} R^2 \right) \left(\frac{v_{SC}}{R} \right)^2 = \tfrac{3}{4} m_{SC} v_{SC}^2,$$

$$v_{SC} = \sqrt{\tfrac{4}{3}} \sqrt{gh}.$$

The solid sphere (SS) produces

$$m_{SS}gh = \tfrac{1}{2}m_{SS}v_{SS}^2 + \tfrac{1}{2}\left(\tfrac{2}{5}m_{SS}R^2\right)\left(\frac{v_{SS}}{R}\right)^2 = \tfrac{7}{10}m_{SS}v_{SS}^2,$$

$$v_{SS} = \sqrt{\tfrac{10}{7}}\sqrt{gh}.$$

The thin-walled hollow sphere (TWHS) gives

$$m_{TWHS}gh = \tfrac{1}{2}m_{TWHS}v_{TWHS}^2 + \tfrac{1}{2}\left(\tfrac{2}{3}m_{TWHS}R^2\right)\left(\frac{v_{TWHS}}{R}\right)^2 = \tfrac{5}{6}m_{TWHS}v_{TWHS}^2,$$

$$v_{TWHS} = \sqrt{\tfrac{6}{5}}\sqrt{gh}.$$

Comparing the leading factors, we see that the solid sphere has the largest velocity, followed by the solid cylinder, the thin-walled hollow sphere, and the thin-walled hollow cylinder.

EVALUATE The largest moment of inertias resulted in the smallest velocities at the bottom of the ramp, since energy went into rotational kinetic energy. We see that the radii and masses cancel in this problem: The velocity at the bottom is dependent only upon the distribution of mass in the object.

In what order do the four objects reach the bottom? Since all the velocities depend on the square root of the height, those with the fastest velocity reach the bottom first.

Problems

1: Constant angular acceleration in a pottery wheel

A pottery wheel is rotating with an initial angular velocity ω_0 when the wheel's drive motor is turned on. The wheel increases to a final angular velocity of 125 rpm while making 30.0 revolutions in 25.0 seconds. Find the initial angular velocity and angular acceleration, assuming that the latter is constant. A pottery wheel is essentially a cylinder rotating about a vertical axis driven by a motor.

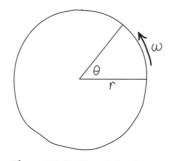

Figure 9.1 Problem 1 sketch.

Solution

IDENTIFY The wheel exhibits constant angular acceleration, so we use the equations for constant angular acceleration to solve the problem. The target variables are the initial angular velocity and the angular acceleration.

SET UP Figure 9.1 shows a sketch of the pottery wheel. For consistency, we'll use radians and seconds as our units and convert the given quantities.

We are given the final angular velocity, the angular displacement, and the time. We need to find the initial angular velocity and the angular acceleration. None of the equations for constant angular acceleration allow us to solve for both unknowns at once, so we'll solve for the initial angular velocity first and then use the results to solve for the angular acceleration.

EXECUTE The angular displacement can be written in terms of the average angular velocity and the time interval as

$$\theta - \theta_0 = \tfrac{1}{2}(\omega_z + \omega_{0z})t.$$

Solving for the initial angular velocity gives

$$\omega_{0z} = \frac{2(\theta - \theta_0)}{t} - \omega_z.$$

The quantity $(\theta - \theta_0)$ is our angular displacement, 30.0 revolutions. The quantity ω_z is the final angular velocity, 125 rpm. We convert the revolutions to radians by multiplying by $(2\pi \text{ rad/rev})$ and the rpm to radians/second by multiplying by $(2\pi \text{ rad/rev})(1 \text{ min/60 s})$. Solving for ω_{0z} gives

$$\omega_{0z} = \frac{2(30.0 \text{ rev})(2\pi \text{ rad/rev})}{25.0 \text{ s}} - (125 \text{ rpm})(2\pi \text{ rad/rev})(1 \text{ min/60 s}) = 1.99 \text{ rad/s}.$$

The initial angular velocity is 1.99 rad/s, or 19.0 rpm. To find the angular acceleration, we use the relationship between angular velocity, angular acceleration, and time:

$$\omega_z = \omega_{0z} + \alpha_z t.$$

Solving for the angular acceleration, we obtain

$$\alpha_z = \frac{\omega_z - \omega_{0z}}{t} = \frac{(125 \text{ rpm})(2\pi \text{ rad/rev})(1 \text{ min/60 s}) - (1.99 \text{ rad/s})}{25.0 \text{ s}} = 0.444 \text{ rad/s}^2.$$

The angular acceleration is 0.444 rad/s^2.

EVALUATE This problem reminds us of the problems involving constant linear acceleration we first encountered in Chapter 2. The same problem-solving strategy applies: Draw a diagram, check for constant acceleration, find one or more equations that can be used to solve for the unknowns, and reflect upon the results. As we've seen here, we may need to use more than one equation for the solution, and we must watch our units carefully.

2: A slowing pottery wheel

A pottery wheel rotating at 30 revolutions per minute is shut off and slows uniformly, coming to a stop in 2 complete revolutions. Find the angular acceleration and the time it takes to come to a stop.

Solution

IDENTIFY The wheel exhibits constant angular acceleration as it slows uniformly, so we use the equations for constant angular acceleration to solve the problem. The target variables are the angular acceleration and the time required to stop.

SET UP We are given the initial angular velocity, the final angular velocity (zero), and the angular displacement for the wheel to come to a stop. We will use two of the equations for constant angular acceleration to solve for the two unknowns.

EXECUTE The angular displacement, velocity, and acceleration are related by the formula

$$\omega_z^2 = \omega_{0z}^2 + 2\alpha_z(\theta - \theta_0).$$

Solving for the angular acceleration gives

$$\alpha_z = \frac{\omega_z^2 - \omega_{0z}^2}{2(\theta - \theta_0)} = \frac{(0)^2 - (30 \text{ rev/min})^2}{2(2 \text{ rev})} = -225 \text{ rev/min}^2 \left(\frac{2\pi \text{ rad}}{\text{rev}}\right)\left(\frac{1 \text{ min}}{60 \text{ s}}\right)^2 = -0.393 \text{ rad/s}^2.$$

To solve for time, we use the acceleration–velocity relation:

$$\omega_z = \omega_{0z} + \alpha_z t.$$

Solving for the time yields

$$t = \frac{\omega_z - \omega_{0z}}{\alpha_z} = \frac{(0) - (30 \text{ rev/min})}{-225 \text{ rev/min}^2} = 0.133 \text{ min} = 8.0 \text{ s}.$$

The wheel slows to a stop at a rate of -0.393 rad/s^2, stopping in 8.0 s.

EVALUATE This problem reminds us of our earlier problems involving constant linear acceleration. You should be able to become proficient at these problems rather quickly. Just make sure that you watch your units!

3: Energy in a wheel–stone system

A thin light string is wrapped around the rim of a spoked wheel that can rotate without friction around its center axle. An 8.00 kg stone is attached to the end of the string as shown in Figure 9.2. If the stone is released from rest, how far does it travel before attaining a speed of 4.80 m/s? The spoked wheel is made of a central hub (a solid uniform cylinder of radius 7.50 cm and mass 22.0 kg) attached to a rim (a thin-walled hollow cylinder of radius 30.0 cm mass 12.0 kg) by spokes of negligible mass.

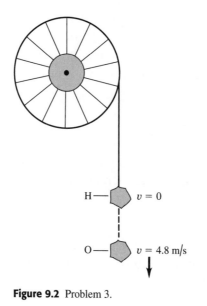

H — $v = 0$

O — $v = 4.8$ m/s

Figure 9.2 Problem 3.

Solution

IDENTIFY No work is done by external forces, so mechanical energy is conserved. The target variable is the height the stone falls.

SET UP Initially, there is only gravitational potential energy. When the stone is released, the gravitational potential energy is transformed into kinetic energy of the stone and wheel. We'll set the origin at the point where the stone reaches a speed of 4.80 m/s; the starting position is a distance H above the origin, as we see in Figure 9.2. We'll find the moment of inertia of the wheel by combining the moment of inertia of a solid cylinder with the moment of inertia of a thin-walled hollow cylinder.

EXECUTE Energy conservation relates the initial and final energies:

$$K_1 + U_1 = K_2 + U_2.$$

Initially, there is only U_{grav}. At the origin, the gravitational potential energy has been transformed into the kinetic energies of the stone and wheel:

$$U_{grav} = K_{stone} + K_{wheel}.$$

Replacing the energies yields

$$m_{stone}gH = \tfrac{1}{2}m_{stone}v^2 + \tfrac{1}{2}I_{wheel}\omega^2.$$

We need to find the moment of inertia of the wheel and its angular velocity when the stone reaches its final velocity. The moment of inertia of the wheel is the algebraic sum of the moment of inertia of the central hub plus the moment of inertia of the outer rim (the thin-walled cylinder.) Using Table 9.2 of the text, we find the total moment of inertia of the wheel:

$$I_{wheel} = I_{solid\ cylinder} + I_{thin\text{-}walled\ cylinder} = \tfrac{1}{2}M_{hub}R_{hub}^2 + M_{rim}R_{rim}^2.$$

The speed of the stone is the tangential speed of the wheel, so we can find the angular speed of the wheel from the equation

$$\omega = \frac{v}{R_{rim}}.$$

Substituting the two expressions we found into the energy relation gives

$$m_{stone}gH = \tfrac{1}{2}m_{stone}v^2 + \tfrac{1}{2}\left(\tfrac{1}{2}M_{hub}R_{hub}^2 + M_{rim}R_{rim}^2\right)\left(\frac{v}{R_{rim}}\right)^2.$$

Solving for H results in

$$H = \frac{1}{m_{stone}g}\left[\tfrac{1}{2}m_{stone}v^2 + \tfrac{1}{2}\left(\tfrac{1}{2}M_{hub}R_{hub}^2 + M_{rim}R_{rim}^2\right)\left(\frac{v}{R_{rim}}\right)^2\right],$$

$$H = \frac{1}{(8.00\ kg)(9.8\ m/s^2)}\left[\tfrac{1}{2}(8.00\ kg)(4.80\ m/s)^2 + \tfrac{1}{2}\left(\tfrac{1}{2}(22.0\ kg)(0.0750\ m)^2 + (12.0\ kg)(0.300\ m)^2\right)\left(\frac{4.80\ m/s}{0.300\ m}\right)^2\right] = 3.04\ m.$$

After the stone falls 3.04 m, its speed will be 4.80 m/s.

EVALUATE If the stone had fallen freely, it would have attained the final speed when $h = v/\sqrt{2g}$ (1.08 m). Why does it take almost three times this distance to reach the final speed? Looking at the energy conservation equation, we see that the energy is shared between the kinetic energies of the stone and wheel. Roughly two-thirds of the energy goes into the kinetic energy of the wheel.

Practice Problem: Repeat the problem with only the inner hub of the wheel. *Answer*: 1.28 m.

4: Energy in a falling cylinder

A thin, light string is wrapped around a solid uniform cylinder of mass M and radius R as shown in Figure 9.3. The string is held stationary and the cylinder is released from rest. What is the cylinder's radius if it reaches an angular speed of 350.0 rpm after it falls 3.00 m?

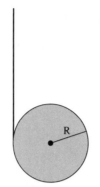

Figure 9.3 Problem 4.

Solution

IDENTIFY Gravity and tension are the only forces acting in this problem, so energy is conserved. The target variable is the cylinder's radius.

SET UP Initially, there is only gravitational potential energy. When the cylinder is released, the gravitational potential energy is transformed into rotational and translational kinetic energy. We'll set the origin 3.00 m below the initial position.

EXECUTE Energy conservation relates the initial and final energies:

$$K_1 + U_1 = K_2 + U_2.$$

Initially, there is only U_{grav}. At the origin, the gravitational potential energy has been transformed into the total kinetic energy of the cylinder:

$$U_{grav} = K_{translational} + K_{rotational}.$$

Replacing the energies, we obtain

$$Mgh = \tfrac{1}{2}Mv^2 + \tfrac{1}{2}I\omega^2.$$

Now, recall that the moment of inertia of a solid uniform cylinder is $\tfrac{1}{2}MR^2$. (See Table 9.2 in the text.) The speed of the cylinder is the tangential speed of the wheel, so we can find the angular speed of the wheel from the relation

$$v = \omega R.$$

Substituting the preceding expressions into the energy relation gives

$$Mgh = \tfrac{1}{2}Mv^2 + \tfrac{1}{2}I\omega^2 = \tfrac{1}{2}M(\omega R)^2 + \tfrac{1}{2}(\tfrac{1}{2}MR^2)\omega^2 = \tfrac{3}{4}M(\omega R)^2.$$

Solving for R, we get

$$R = \frac{\sqrt{\tfrac{4}{3}gh}}{\omega} = \frac{\sqrt{\tfrac{4}{3}(9.80 \text{ m/s}^2)(3.00 \text{ m})}}{(350 \text{ rpm})(2\pi \text{ rad/rev})(1 \text{ min/60 s})} = 0.171 \text{ m}.$$

The cylinder's radius is 17.1 cm.

EVALUATE We see that the mass of the cylinder cancels in the conservation-of-energy equation; the results apply to a cylinder of any mass.

Practice Problem: Repeat the problem with a thin hoop replacing the cylinder. *Answer*: 14.8 cm.

5: Cylinder rolling down a ramp

A uniform hollow cylinder rolls along a horizontal floor and then up a flat ramp without slipping. The ramp is inclined at 20.0°. How far along the ramp does the cylinder roll before stopping if its initial forward speed is 12.0 m/s? The hollow cylinder has a mass of 6.5 kg, an inner radius of 0.13 m, and an outer radius of 0.25 m.

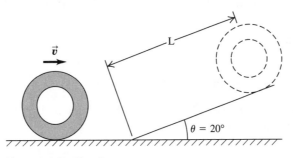

Figure 9.4 Problem 5.

Solution

IDENTIFY There are no nonconservative forces (ignoring air drag), so we use energy conservation. The target variable is the distance up the ramp the cylinder travels before stopping momentarily.

SET UP Figure 9.4 shows a sketch of the problem. Initially, the cylinder has translational and rotational kinetic energy. When the cylinder stops momentarily at the top of the ramp, the kinetic energy has been totally transformed to gravitational potential energy. We'll set the origin at the base of the ramp.

EXECUTE Energy conservation relates the initial and final energies:

$$K_1 + U_1 = K_2 + U_2.$$

Initially, the cylinder has only kinetic energy. At its highest point, the cylinder has only gravitational potential energy. Thus,

$$K_{\text{translational}} + K_{\text{rotational}} + 0 = 0 + U_{\text{grav}}.$$

Replacing the energies with their equivalent expressions yields

$$\tfrac{1}{2}Mv^2 + \tfrac{1}{2}I\omega^2 = Mgy.$$

Recall that the moment of inertia of a hollow uniform cylinder is $\tfrac{1}{2}M(R_1^2 + R_2^2)$. (See Table 9.2 in the text.) The speed of the cylinder is the tangential speed of the wheel at the outer radius, so we can replace the speed with

$$\omega = \frac{v}{R_2}.$$

Substituting these expressions into the energy relation gives

$$\tfrac{1}{2}Mv^2 + \tfrac{1}{2}I\omega^2 = \tfrac{1}{2}Mv^2 + \tfrac{1}{2}\left(\tfrac{1}{2}M(R_1^2 + R_2^2)\right)\left(\frac{v}{R_2}\right)^2 = Mgy.$$

Solving for y, the maximum vertical height, we obtain

$$y = \frac{v^2}{g}\left[\frac{1}{2} + \frac{1}{4}\left(\frac{R_1^2 + R_2^2}{R_2^2}\right)\right] = \frac{(12.0 \text{ m/s})^2}{(9.8 \text{ m/s}^2)}\left[\frac{1}{2} + \frac{1}{4}\left(\frac{(0.13 \text{ m})^2 + (0.25 \text{ m})^2}{(0.25 \text{ m})^2}\right)\right] = 12.0 \text{ m}.$$

The cylinder's maximum vertical height is 12.0 m. To find the distance along the ramp, we use the sine relation:

$$L = \frac{y}{\sin 20°} = \frac{(12.0 \text{ m})}{\sin 20°} = 35.1 \text{ m}.$$

The hollow cylinder rolls 35.1 m up along the ramp.

EVALUATE How far up the ramp would the cylinder travel without friction? It would move 21.5 m along the ramp without friction. This result shows how the added initial rotational kinetic energy results in a greater distance along the ramp.

6: Moment of inertia of a rectangular sheet

Find the moment of inertia of a uniform thin rectangular sheet of metal with mass M, length L, and width W about the x-axis in Figure 9.5.

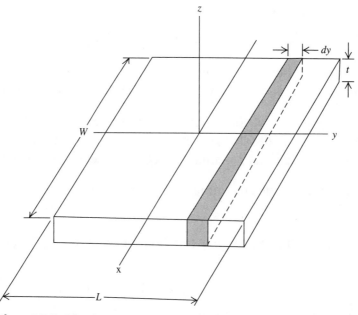

Figure 9.5 Problem 6.

Solution

IDENTIFY The sheet is a continuous distribution of mass, so we must integrate to find the moment of inertia. We break the sheet up into thin strips along the x-axis as shown. The target variable is the moment of inertia.

SET UP The mass density of the sheet is the total mass divided by the volume, or

$$\rho = \frac{M}{WLt},$$

where t is the thickness of the sheet. The volume of the thin strip along the x-axis is

$$dV = Wt\,dy.$$

The mass of the strip is the density multiplied by the volume. We will integrate the product of the mass and r^2 to find the moment of inertia.

EXECUTE The moment of inertia is given by

$$I = \int r^2\,dm.$$

We will integrate from $-L/2$ to $+L/2$, since the sheet is centered on the origin. Replacing dm and adding the limits of integration gives

$$\begin{aligned}
I &= \int_{-L/2}^{L/2} r^2 \rho\,dV \\
&= \int_{-L/2}^{L/2} y^2 \frac{M}{WLt} Wt\,dy \\
&= \frac{M}{L} \int_{-L/2}^{L/2} y^2\,dy \\
&= \frac{M}{L} \frac{y^3}{3}\Big|_{-L/2}^{L/2} = \frac{M}{L}\left[\frac{(L/2)^3}{3} - \frac{(-L/2)^3}{3}\right] \\
&= \frac{ML^2}{12}.
\end{aligned}$$

The moment of inertia of the sheet rotating about the x-axis is $ML^2/12$.

EVALUATE We see that the moment of inertia does not depend on the width or thickness of the sheet when it is rotated about the x-axis. If we look at other shapes in Table 9.2, we see that neither does the moment of inertia depend upon dimensions along the axis of rotation.

Practice Problem: Use the parallel-axis theorem to check that this result agrees with the moment of inertia of a thin rectangular plate rotated about the edge.

Try It Yourself!

1: Motion of flywheel

A flywheel is a disk-shaped mass that rotates about its central perpendicular axis. A flywheel 1.0 m in diameter rotates with an initial velocity of 500 rpm. It increases its speed to 1000 rpm in 20.0 s. Assuming constant acceleration, find the angular acceleration and the angular displacement of the flywheel as it increases its speed from 500 to 1000 rpm.

Solution Checkpoints

IDENTIFY AND SET UP Determine the target variables and identify the appropriate constant-angular-acceleration equations needed to find the target variables.

EXECUTE The angular acceleration is found from the relation

$$\omega_z = \omega_{0z} + \alpha_z t.$$

This gives an angular acceleration of 1510 rev/min². The angular displacement is found from the formula

$$\omega_z^2 = \omega_{0z}^2 + 2\alpha_z(\theta - \theta_0).$$

This gives an angular displacement of 248 rev.

EVALUATE We see that this problem closely parallels the linear kinematics problems we have seen throughout the previous chapters. We will continue to use angular kinematics as we investigate rotations.

2: Energy in a grinding wheel

How much energy is dissipated when a 2.0 kg grinding wheel of radius 0.1 m is brought to rest from an initial velocity of 3000 rpm? What is the average power dissipated if the wheel stops in 10 rev? Assume constant angular acceleration.

Solution Checkpoints

IDENTIFY AND SET UP The grinding wheel loses all of its energy as it stops, so you must find the initial kinetic energy. The moment of inertia of the wheel is that of a disk. To find the power, you must find the time it takes the wheel to stop and divide the energy by the time. The target variables are the energy and the power.

EXECUTE The initial kinetic energy of the grinding wheel is given by

$$K = \tfrac{1}{4}MR^2\omega^2.$$

The final kinetic energy is zero, so the wheel loses 493 J.

We find the time it takes the wheel to stop by directly combining two kinematics equations:

$$\omega_z^2 = \omega_{0z}^2 + 2\alpha_z(\theta - \theta_0) \text{ and}$$
$$\omega_z = \omega_{0z} + \alpha_z t.$$

These equations result in a time of 0.40 s. The power is then the energy lost divided by the time, giving 1.23 kW.

EVALUATE This problem combines energy, kinematics, and power. How can we best check our results?

10 Dynamics of Rotational Motion

Summary

In this chapter, we will investigate the dynamics of rotational motion to learn what gives an object angular acceleration. We will define torque—the turning or twisting effort of a force—and learn how to apply it to both equilibrium and nonequilibrium situations. Work and power for rotating systems will also be investigated. Angular momentum will be introduced and become the basis of an important new conservation law that will lead to an analysis of a spinning gyroscope and the motion called precession. The linear dynamics foundation we have developed throughout the text will help build our intuition about rotational dynamics.

Objectives

After studying this chapter, you will understand

- The definition and meaning of torque.
- How to identify torques acting on a body.
- The equation of motion for rotational systems and how to apply it to problems.
- How to apply work and power to rotational dynamics problems.
- The definition of angular momentum and how it changes with time.
- How to apply conservation of angular momentum to problems.
- The concept of precession as it applies to gyroscopes.

Concepts and Equations

Term	Description
Torque	Torque is the tendency of a force to cause or change rotational motion about a chosen axis. The magnitude of torque is the magnitude of the force (F) times the moment arm (l), which is the perpendicular distance between the axis and the line of force: $$\tau = Fl.$$ For a force $\vec{F}$ applied at point O and a vector $\vec{r}$ from the chosen axis to point O, the torque is given by $$\vec{\tau} = \vec{r} \times \vec{F}.$$ The SI unit of torque is the newton-meter (Nm).
Combined Translation and Rotation	The motion of a rigid body, moving through space and rotating, can be regarded as translational motion of the center of mass plus rotational motion about the center of mass. The kinetic energy, net force, and net torque are given respectively, by $$K = \tfrac{1}{2}Mv_{cm}^2 + \tfrac{1}{2}I_{cm}\omega^2,$$ $$\sum \vec{F}_{ext} = M\vec{a}_{cm},$$ $$\sum \vec{\tau}_z = I_{cm}\alpha_z.$$ In the case of rolling without slipping, the motion of the center of mass is related to the angular velocity by $$v_{cm} = R\omega.$$
Work Done by a Torque	The work done by a torque is given by $$W = \int_{\theta_1}^{\theta_2} \tau_z d\theta.$$ For a constant torque, $$W = \tau_z \Delta\theta.$$ The power provided by a torque is the product of the torque and the angular velocity of the body: $$P = \tau_z \omega_z.$$
Angular Momentum	The angular momentum L of a particle with respect to point O is the vector product of the position vector and momentum of the particle: $$\vec{L} = \vec{r} \times \vec{p}.$$ The angular momentum of a symmetrical body rotating about a stationary axis of symmetry is $$\vec{L} = I\vec{\omega}.$$ Our convention is that counterclockwise rotations have positive L and clockwise rotations have negative L.
Rotational Dynamics and Angular Momentum	The net external torque on a system is equal to the rate of change of the angular momentum of the system: $$\sum \vec{\tau} = \frac{d\vec{L}}{dt}.$$ If the net external torque acting on a system is zero, the total angular momentum is conserved and remains constant.

Conceptual Questions

1: Ranking torques

In the following diagram, each rod pivots about the indicated axis with the indicated force:

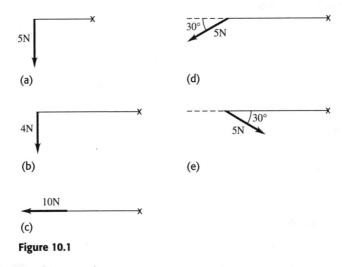

Figure 10.1

Rank the diagrams in order of increasing torque.

Solution

IDENTIFY, SET UP, AND EXECUTE Torque is given by $\tau = rF\sin\phi$, so we must examine the magnitude of the force, the location at which the force is applied, and the direction in which the force is applied. Comparing (a) and (b), we see that both forces act in the same direction, the force is larger in (a), and the moment arm is larger in (b). We estimate the moment arm in (a) as half the moment arm of (b). Since the force in (a) is less than double the force in (b), the torque in (b) is greater than that in (a). No torque is generated in (c), because the force acts along the moment arm. The force and moment arm are the same in (d) and (e), but the directions are different. However, since both forces act 30° from the horizontal, the components of the forces perpendicular to the moment arm are the same for both and the torque in (d) and (e) are the same. The vertical component of the force in (d) and (e) is 2.5 N, which is less than the force in (b) and less than half of the force in (a). Diagrams (d) and (e) both show less torque than diagrams (a) and (b).

Ranked in order of increasing torque, the diagrams are therefore (c), (d) = (e), (a), (b).

EVALUATE Torque is the most complicated quantity that we have discussed thus far. It depends on the magnitude and direction of the force responsible for it, as well as where that force acts relative to the rotation axis. Gaining intuition about torque will help guide you through problems.

2: Massless versus massive pulleys

Why were massless pulleys used in problems from the previous chapters?

Solution

IDENTIFY, SET UP, AND EXECUTE Consider the net torque acting on a pulley with a rope resting on the pulley. The left segment of rope has a tension T_L, and the right segment of rope has a tension T_R. The net torque is then

$$\sum \tau = T_R R - T_L R = I\alpha.$$

Each tension creates a torque *TR,* since the rope will be perpendicular to the radius and the two torques are opposite in direction. Massless pulleys have no moment of inertia (since their mass is zero); therefore, the net torque acting on the pulley is zero. If the net torque is zero, the torques in each segment must be equal and the tensions are equal in each rope.

When we include the pulley's mass, the pulley has a moment of inertia. If the pulley accelerates, there must be a difference in the left and right torques and the tensions must also be different. Only when the angular acceleration is zero are the two tensions equal.

Massless pulleys were used in earlier chapters to avoid having to include torque in our analysis. Using massless pulleys simplified our analysis and let us focus on learning about forces.

EVALUATE From now on, we must assume that the tensions in segments of rope may vary and we must identify each segment's tension separately.

How does the acceleration of each segment of rope compare when we include a pulley's mass? There is no change: Objects connected by the rope are constrained to have the same magnitude of acceleration.

3: Spinning on a roundabout

You are standing at the center of a rotating playground roundabout (a round, horizontal plate that spins about its center axis). As you move to the edge, will your angular speed increase, decrease, or stay the same?

Solution

IDENTIFY, SET UP, AND EXECUTE To simplify our analysis, we ignore friction in the roundabout. Without friction, there is no torque to slow the roundabout, so angular momentum is conserved. As you move to the edge, the moment of inertia of the system increases. (Your mass moves to a greater radius.) For angular momentum to be conserved, the angular speed must be reduced.

EVALUATE Like linear momentum and energy, angular momentum is a conserved quantity. We could solve this problem numerically by picking initial and final angular momenta before and after you moved and setting them equal to each other.

4: Standing on a turntable

You are standing on a small frictionless turntable with your arms outstretched and spinning about the axis of the turntable. As you pull your arms in, you spin faster. Does your rotational kinetic energy increase, decrease, or stay the same?

Solution

IDENTIFY AND SET UP There are no external torques acting on you or the turntable, so angular momentum is conserved. You spin faster, since you decrease your moment of inertia as you bring your arms in, leading to a greater angular velocity. Rotational kinetic energy is given by

$$K = \tfrac{1}{2}I\omega^2.$$

Since kinetic energy depends on both moment of inertia and angular velocity, and since both change, we'll have to consider more information. Angular momentum is conserved, so $I\omega$ is constant.

EXECUTE We include angular momentum in the kinetic energy:

$$K = \tfrac{1}{2}L\omega.$$

Since L is constant and ω increases as you bring your arms in, the kinetic energy increases.

EVALUATE Where does the increased energy come from? You must do work to pull your arms in; doing work on the system increases its kinetic energy.

This example also applies to a figure skater spinning on ice.

Problems

1: Balancing a food tray

A waiter balances a tray of food on his hand. On the tray is a 0.40 kg drink and a 2.0 kg lobster dinner. The drink is placed 6.5 cm from one edge of the tray, and the lobster dinner is placed 8.0 cm from the opposite edge. The tray has a mass of 1.2 kg and a diameter of 42 cm. Where should the waiter hold the tray so that it doesn't tip over?

Solution

IDENTIFY Figure 10.2 shows a sketch, and a free-body diagram, of the food tray. The tray should be held so that it is in equilibrium and the net torque on it is zero. The target variable is the location at which the waiter's hand holds the tray.

SET UP The forces on the food tray include the force of the waiter's hand holding the tray, the weight of the tray, and the normal forces due to the lobster dinner and drink. To find the location of the hand, we need to consider the torque acting on the tray. When the tray is in equilibrium, the net torque must be zero about any axis. We'll take the axis to be the left edge, marked by an X.

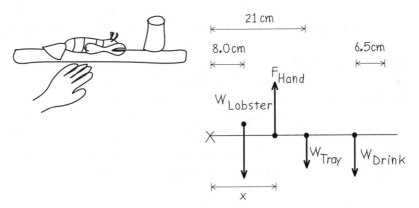

Figure 10.2 Problem 1 sketch and free-body diagram.

When we choose the left edge as the axis, we note that four torques act on the tray, corresponding to the four forces on the tray. Each of the four forces is applied perpendicular to the moment arm (the plane of the tray); each torque is the magnitude of the force times the distance from the axis. We'll take counterclockwise torques to be positive.

EXECUTE Since the tray is in equilibrium, the net torque is zero:

$$\sum \tau = \tau_{\text{tray}} + \tau_{\text{lobster}} + \tau_{\text{drink}} + \tau_{\text{hand}} = 0.$$

Writing the four torques explicitly, we have

$$\sum \tau = \tau_{\text{tray}} + \tau_{\text{lobster}} + \tau_{\text{drink}} + \tau_{\text{hand}} = -m_{\text{tray}}gx_{\text{tray}} - n_{\text{lobster}}x_{\text{lobster}} - n_{\text{drink}}x_{\text{drink}} + n_{\text{hand}}x_{\text{hand}} = 0.$$

The first term is the torque due to the weight of the tray; the moment arm is half the tray diameter (at the center of mass). The second and third terms are the torques due to the normal forces of the lobster dinner and drink, respectively. The normal force is equal to the weight of the objects, and the moment arm is the distance from the left edge of the tray. The first three terms are negative, since they are all in the clockwise direction. The last term is the torque due to the normal force of the waiter's hand and is the only positive term, because it is counterclockwise. We need to find this normal force of his hand to solve the problem. We find the force by using Newton's first law:

$$\sum F = 0.$$

In the vertical direction, four forces act on the tray. We have

$$\sum F = -m_{tray}g - n_{lobster} - n_{drink} + n_{hand} = 0.$$

Solving yields

$$n_{hand} = (m_{tray} + m_{lobster} + m_{drink})g = ((1.2 \text{ kg}) + (2.0 \text{ kg}) + (0.40 \text{ kg}))(9.8 \text{ m/s}^2) = 35.3 \text{ N}.$$

We can now solve for the location of the waiter's hand:

$$x_{hand} = \frac{m_{tray}gx_{tray} + m_{lobster}gx_{lobster} + m_{drink}gx_{drink}}{n_{hand}}$$

$$= \frac{((1.2 \text{ kg})(0.21 \text{ m}) + (2.0 \text{ kg})(0.080 \text{ m}) + (0.40 \text{ kg})(0.42 \text{ m} - 0.065 \text{ m}))(9.8 \text{ m/s}^2)}{(35.3 \text{ N})}$$

$$= 0.15 \text{ m}.$$

The waiter should hold the tray 15 cm from the left edge (closest to the lobster dinner).

EVALUATE This equilibrium problem required us to apply both the equilibrium torque and the equilibrium force conditions to solve the problem. You should be familiar with the equilibrium force condition and should need to gain expertise only in torque to solve similar problems.

Experience shows that we may be able to simplify similar problems by picking an axis that coincides with the location at which a force is applied. This choice will reduce the number of torques in the problem. In the current problem, for example, we could have chosen the axis to be at the location of the lobster dinner, thus removing the torque due to the normal force of the lobster dinner.

2: Tension in string attached to a falling cylinder

A thin, light string is wrapped around the outer rim of a uniform hollow cylinder of mass 12.0 kg, inner radius 15.0 cm, and outer radius 30.0 cm. The cylinder is released from rest. What is the tension in the string as the cylinder falls?

Solution

IDENTIFY We will apply Newton's second law and its rotational analog to solve for the tension, the target variable.

SET UP Figure 10.3 shows a sketch, and a free-body diagram, of the situation. As the cylinder falls, it will accelerate downward and rotate about its central axis. In falling, the cylinder will rotate faster and undergo angular acceleration. The cylinder has both a net force and a net torque acting on it.

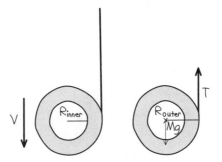

Figure 10.3 Problem 2 sketch and free-body diagram.

EXECUTE We first apply Newton's second law to the translational motion of the center of mass in the vertical direction. The only forces acting in the vertical direction are gravity and tension. We have

$$\sum F_y = Mg - T = Ma_{cm,y}.$$

The moment of inertia of the hollow cylinder is $I = \frac{1}{2}M(R_{inner}^2 + R_{outer}^2)$. The one torque acting on the cylinder as it rotates about its central axis is due to the tension force. Gravity acts on the center of mass, but creates no torque about the central axis. The torque acts perpendicular to the outer radius. Thus,

$$\sum \tau_z = TR_{outer} = I\alpha_z = \frac{1}{2}M(R_{inner}^2 + R_{outer}^2)\alpha_z.$$

We can relate the two accelerations, since the cylinder falls without slipping:

$$a_y = R_{outer}\alpha_z.$$

Solving for the acceleration of the center of mass in the first equation yields

$$a_y = \frac{Mg - T}{M}.$$

Substituting the last two equations into the torque result gives

$$TR_{outer} = \frac{1}{2}M(R_{inner}^2 + R_{outer}^2)\frac{a_y}{R_{outer}} = \frac{1}{2}M(R_{inner}^2 + R_{outer}^2)\frac{Mg - T}{MR_{outer}},$$

$$T = \frac{(R_{inner}^2 + R_{outer}^2)Mg}{R_{inner}^2 + 3R_{outer}^2} = \frac{((15.0 \text{ cm})^2 + (30.0 \text{ cm})^2)(12.0 \text{ kg})(9.8 \text{ m/s}^2)}{(15.0 \text{ cm})^2 + 3(30.0 \text{ cm})^2} = 45.2 \text{ N}.$$

The tension in the string is 45.2 N.

EVALUATE This problem resembles our earlier Newton's-law problems, but with the addition of torque. Once torque is included, the problem becomes a relatively straightforward algebraic one. We also see that the tension in the string is less than that for a stationary cylinder. If the cylinder were stationary, the tension would have been 118 N.

Practice Problem: At what rate does the cylinder accelerate? *Answer*: 6.03 m/s².

3: Acceleration and tension of two blocks connected by a pulley

Two blocks are connected to each other by a light cord passing over a pulley as shown in Figure 10.4. Block *A* has a mass of 5.00 kg and block *B* has a mass of 4.00 kg. The pulley has a mass of 8.00 kg and a radius of 4.00 cm. Find the acceleration of the blocks and the tensions in the horizontal and vertical segments of the cord. Assume that the pulley is a solid, uniform disk and there is no friction between block *A* and the table.

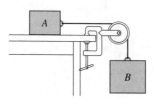

Figure 10.4 Problem 3.

Solution

IDENTIFY We'll apply the net-force and net-torque equations to solve the problem. The accelerations of both blocks are the same, as we saw in Chapter 5. Our target variables are the two tensions and the acceleration of the blocks.

SET UP Figure 10.5 shows the free-body diagram of the two blocks and the pulley. The forces on the blocks include tension, gravity, and the normal force (block *A*). We assume that the tensions of the two segments are not equal and label them T_A and T_B. The two tension forces lead to two torques acting on the pulley. (The axis of rotation is the center of the pulley.)

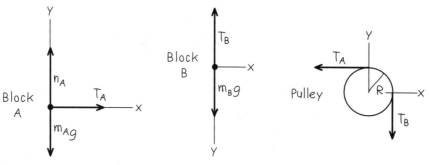

Figure 10.5 Problem 3 free-body diagrams.

EXECUTE We first apply Newton's second law to each block. Block *A* (with mass m_A) accelerates in the *x* direction due to tension T_A, so

$$\sum F_x = T_A = m_A a.$$

Gravity and tension T_B act on block *B* (with mass m_B) and that block accelerates at the same rate as block *A* in the *y* direction, so

$$\sum F_y = m_B g + (-T_B) = m_B a.$$

As block *B* falls, the pulley's rotational speed increases. The net torque on the pulley is

$$\sum \tau_z = \tau_A - \tau_B = I\alpha_z,$$

where we have taken the counterclockwise torque as positive and clockwise torque as negative. The moment of inertia of a uniform cylinder is $I = \frac{1}{2}MR^2$. We assume that the cord doesn't slip on the pulley, so we relate the angular acceleration of the pulley to the tangential acceleration of the cord (a):

$$\alpha_z = -\frac{a}{R}.$$

We included a minus sign in the equation because the pulley rotates clockwise (negative, according to our convention). The tension forces act perpendicular to the moment arm, so the torques are simply TR. Rewriting the net torque, we have

$$\sum \tau_z = \tau_A - \tau_B = T_A R - T_B R = I\alpha_z = \frac{1}{2}MR^2\left(-\frac{a}{R}\right).$$

Simpifying yields

$$T_B - T_A = \frac{1}{2}Ma.$$

Our second-law equations are used to replace the tensions:

$$m_B g - m_B a - m_A a = \frac{1}{2}Ma.$$

Solving for the acceleration gives

$$a = \frac{m_B g}{m_B + m_A + \frac{1}{2}M} = \frac{(4.00\,\text{kg})(9.8\,\text{m/s}^2)}{(4.00\,\text{kg}) + (5.00\,\text{kg}) + \frac{1}{2}(8.00\,\text{kg})} = 3.02\,\text{m/s}^2.$$

Using this result to find the two tensions, we get

$$T_A = m_A a = (5.00\,\text{kg})(3.02\,\text{m/s}^2) = 15.1\,\text{N},$$
$$T_B = m_B g - m_B a = (4.00\,\text{kg})(9.80\,\text{m/s}^2) - (4.00\,\text{kg})(3.02\,\text{m/s}^2) = 27.1\,\text{N}.$$

The blocks accelerate at $3.02\,\text{m/s}^2$, the tension in the horizontal segment of the cord is $15.1\,\text{N}$, and the tension in the vertical segment of the cord is $27.1\,\text{N}$.

EVALUATE The tensions in the two segments of the cord differ by almost a factor of two. Recall from our problems in Chapter 5 that the tension was constant in both segments of the cord. What causes this difference? The current problem includes the pulley's mass, resulting in some energy spent on increasing the pulley's angular velocity, leaving less energy available for the blocks. This problem also illustrates why we let the first pulleys we encountered be massless.

4: Solid cylinder rolling down ramp

A solid cylinder rolls without slipping down an incline of 40°. Find the acceleration and minimum coefficient of friction needed to prevent slipping.

Solution

IDENTIFY We'll apply translational and rotational dynamics to the cylinder. Since the cylinder doesn't slip, we will use the relationship between the linear and angular acceleration of the cylinder. The target variables are the acceleration and coefficient of friction.

SET UP Figure 10.6 shows a sketch, and a free-body diagram, of the cylinder on the incline. Gravity, the normal force, and the frictional force act on the cylinder. If we set the axis of rotation at the center of the cylinder, a torque due to friction acts on the cylinder. We'll apply the net-force and net-torque equations to solve the problem.

A rotated coordinate system is included in the free-body diagram to simplify our analysis.

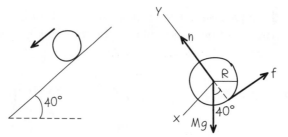

Figure 10.6 Problem 4 sketch and free-body diagram.

EXECUTE We first apply Newton's second law to the translational motion along the x-axis:

$$\sum F_x = Mg\sin\theta - f_s = Ma_x.$$

The equation of motion for the rotation about the axis is

$$\sum \tau_z = f_s R = I_{cm}\alpha_z = \tfrac{1}{2}MR^2\alpha_z,$$

where we included the moment of inertia $(I = \tfrac{1}{2}MR^2.)$ The translational and rotational accelerations are related by $a_{cm} = \alpha_z R$, since the cylinder rolls without slipping. Combining and writing the second equation in terms of f_s yields

$$f_s = \tfrac{1}{2}Ma_x.$$

We use this result in the first equation and solve for the acceleration:

$$Mg\sin\theta - \tfrac{1}{2}Ma_x = Ma_x,$$
$$a_x = \tfrac{2}{3}g\sin\theta = \tfrac{2}{3}(9.80 \text{ m/s}^2)\sin 40° = 4.20 \text{ m/s}^2.$$

To find the minimum coefficient of static friction, we use the equilibrium equation along the y-axis:

$$\sum F_y = n - mg\cos\theta = 0.$$

The friction force is then

$$f_s = \mu_s n = \mu_s mg\cos\theta = \tfrac{1}{2}Ma_x.$$

Solving for μ_s gives

$$\mu_s = \frac{\tfrac{1}{2}a_x}{g\cos\theta} = \frac{\tfrac{1}{2}(4.20 \text{ m/s}^2)}{(9.8 \text{ m/s}^2)\cos 40°} = 0.280.$$

The cylinder accelerates at 4.20 m/s^2, and the minimum coefficient of static friction that will prevent slipping is 0.280.

Evaluate This solution is a straightforward application of net-force and net-torque problem-solving techniques. We see that neither the mass nor the radius of the cylinder affect the results, which are valid for *any* cylinder rolling down a 40° incline.

5: Pivoting rod

A rod of mass 1.0 kg and length 0.50 m is connected to the ceiling by a frictionless hinge. It is released from rest when it is in the horizontal position. What is its angular velocity when it is vertical?

Solution

IDENTIFY There are no dissipative forces, so we will use conservation of energy. The rod begins with only gravitational potential energy, which is subsequently transformed to kinetic energy. The target variable is the rod's angular velocity when the rod is vertical.

SET UP Figure 10.7 shows a diagram of the rod as it falls. The center of mass of the rod is at the center of the rod; the gravitational potential energy depends on the position of the center of mass. As the rod falls, its rotational kinetic energy increases.

We set the origin of the coordinate axis at the center of the rod when it is in the vertical position. The initial height of the rod in this coordinate system is $h = L/2$.

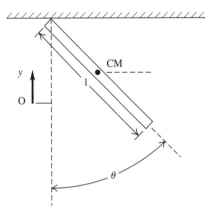

Figure 10.7 Problem 5 sketch and free-body diagram.

EXECUTE Conservation of energy gives

$$K_1 + U_1 = K_2 + U_2.$$

Initially, there is only gravitational potential energy $(MgL/2)$. When the rod is in the vertical position, there is only kinetic energy. The kinetic energy of the rod is

$$K = \tfrac{1}{2}I\omega_z^2 = \tfrac{1}{2}\left(\tfrac{1}{3}ML^2\right)\omega_z^2.$$

Combining these energies, we find that

$$U_1 = K_2,$$

$$Mg\frac{L}{2} = \tfrac{1}{6}ML^2\omega_z^2.$$

Solving for angular velocity gives

$$\omega_z = \sqrt{\frac{3g}{L}} = \sqrt{\frac{3(9.8 \text{ m/s}^2)}{(0.5 \text{ m})}} = 7.67 \text{ rad/s}.$$

The angular velocity of the rod when it is vertical is 7.67 rad/s.

EVALUATE Could we have obtained this result by using rotational dynamics? We could have, although it would have required additional calculations. The forces and torques change with position; hence, the solution would have been more challenging.

6: Baggage carousel

A baggage carousel has a mass of 500.0 kg and can be approximated as a disk of radius 2.0 m. It is rotating freely at an angular velocity of 1.0 rad/s when 10 pieces of baggage, each with a mass of 20.0 kg, are dropped on the edge of the carousel. Assuming that no external torques act on the system, what is the final angular velocity of the system?

Solution

IDENTIFY Since there are no external torques, the total angular momentum of the system remains constant. We will find the initial angular momentum and set it equal to the final angular momentum. The target variable is the carousel's angular velocity after the baggage is added.

SET UP The initial angular momentum is that of the carousel. As the bags are added, they share the angular momentum, resulting in a slower final angular velocity.

EXECUTE The initial angular momentum is

$$L_1 = I_{carousel}\omega_1 = \left(\tfrac{1}{2}M_{carousel}R^2_{carousel}\right)\omega_1.$$

The final angular momentum will be the angular momentum of the carousel plus the angular momentum of the 10 bags:

$$L_2 = I_{carousel}\omega_2 + 10m_{bag}r^2_{bag}\omega_2 = \left(\tfrac{1}{2}M_{carousel}R^2_{carousel}\right)\omega_2 + 10m_{bag}r^2_{bag}\omega_2.$$

Equating the initial and final angular momenta gives

$$\left(\tfrac{1}{2}M_{carousel}R^2_{carousel}\right)\omega_1 = \left(\tfrac{1}{2}M_{carousel}R^2_{carousel}\right)\omega_2 + 10m_{bag}r^2_{bag}\omega_2.$$

All of the radii are 2.0 m, so they cancel. Solving for the final angular velocity gives

$$\omega_2 = \frac{\left(\tfrac{1}{2}M_{carousel}\right)\omega_1}{\left(\tfrac{1}{2}M_{carousel}\right) + 10m_{bag}} = \frac{\left(\tfrac{1}{2}(500.0\text{ kg})\right)(1.0\text{ rad/s})}{\left(\tfrac{1}{2}(500.0\text{ kg})\right) + 10(20.0\text{ kg})} = 0.556\text{ rad/s}.$$

The final angular velocity of the system is 0.556 rad/s.

EVALUATE We see how to apply conservation of angular momentum in this problem. Was energy conserved? Even though there were no external torques, there must have been external forces, since energy was not conserved.

Practice Problem: How much energy was lost as the baggage was added? *Answer*: 218 J.

7: Rotating mass on a string

A 0.10 kg block of mass is attached to a cord that passes through a hole in a horizontal frictionless surface. The block initially is rotating in a circle of radius 0.20 m at an angular velocity of 7.0 rad/s. A force is applied to the cord, shortening it to 0.10 m. What is the new angular velocity of the block?

Solution

IDENTIFY AND SET UP There is no external torque, as the force exerts no torque at the hole. Therefore, the total angular momentum of the system remains constant. The target variable is the block's final angular velocity.

EXECUTE The initial angular momentum is

$$L_1 = I_1\omega_1 = mr_1^2\omega_1.$$

The final angular momentum is

$$L_2 = I_2\omega_2 = mr_2^2\omega_2.$$

Setting these equal to each other gives

$$mr_1^2\omega_1 = mr_2^2\omega_2.$$

Solving results in

$$\omega_2 = \frac{r_1^2}{r_2^2}\omega_1 = 4(7.0 \text{ rad/s}) = 28 \text{ rad/s}.$$

The final angular velocity of the block is 28 rad/s.

EVALUATE We see that the final angular velocity is greater than the initial angular velocity, since the radius decreased. The final result does not depend on the mass of the block. We can go on to find the amount of work done by the force by comparing the initial and final kinetic energies.

Practice Problem: How much work did the force do when shortening the cord? *Answer*: 0.29 J.

Try It Yourself!

1: Torque in a grinding wheel

How much torque is required to bring a 2.0 kg grinding wheel of radius 0.1 m to rest from an initial velocity of 3000 rpm? The grinding wheel stops in 10 rev. How much work is done by the torque to bring the grinding wheel to a halt? Assume constant angular acceleration.

Solution

IDENTIFY AND SET UP We apply our results from Try It Yourself! Problem 9.2 as a starting point. We begin by finding the angular acceleration and use that in combination with the moment of inertia to find the torque. The work is the torque times the angular displacement.

EXECUTE The angular acceleration is -4.5×10^5 rev/min^2, which, when combined with the moment of inertia, gives a torque of -7.85 Nm.
   The work done by the torque is -493 J.

EVALUATE Why are the angular acceleration, torque, and work done all negative values? The negative sign indicates that the grinding wheel is slowing.

2: Mass on a flywheel

A cord is wrapped around the rim of a uniform flywheel of radius 02.0 m and mass 10.0 kg. A 10.0 kg mass is suspended from the cord 10.0 m above the floor. How much time does it take the mass to hit the floor? What is the tension in the rope as it falls?

Solution

IDENTIFY AND SET UP Apply the net-force and net-torque equations to solve the problem. The length of cord pulled from the flywheel is equal to the arc length $r\theta$ at the wheel.

EXECUTE Combining the torque and force equations yields the relation

$$a = \frac{g}{1 + \dfrac{m_{\text{flywheel}}}{2m_{\text{mass}}}}$$

for the acceleration. This result can be combined with linear kinematics to find the time to fall, 1.74 s. The tension in the rope is found from the net-force equation:

$$m_{\text{mass}}g - T = m_{\text{mass}}a.$$

The tension is 32.7 N.

EVALUATE Can energy conservation be used to check the results?

3: Man on a turntable

A turntable with moment of inertia of 2000 kgm^2 makes one revolution every 5.0 s. A man of mass 100 kg standing at the center of the turntable runs out along a radius fixed on the turntable. What is the angular velocity of the turntable when the man is 3.0 m from the center?

Solution

IDENTIFY AND SET UP There are no external torques acting on the system, so angular momentum is conserved. As the man runs out, he changes the angular momentum of the system.

EXECUTE Conservation of angular momentum gives

$$I_{\text{turntable}}\omega_1 = I_{\text{turntable}}\omega_2 + MR^2\omega_2.$$

This results in a final angular velocity of 0.14 rev/s, or one revolution every 7.25 s.

EVALUATE Does the man do positive or negative work on the system? Work is done on the man as he runs out.

Problem Summary

These first 10 chapters of the book form the basis of kinematic and dynamic problem-solving techniques. Future chapters will expand to include additional forces and forms of energy, but still utilize the same problem-solving techniques. Our problem-solving methodology continues to encompass the following techniques:

- Identifying the general procedure to find the solution.
- Sketching the situation when no figure is provided.
- Identifying the forces and torques acting on the bodies.
- Identifying the forms of energy included in the problem.
- Drawing free-body diagrams of the bodies.
- Applying appropriate coordinate systems to the diagrams.
- Applying the equations of motion to find relations among the forces, masses, and accelerations.
- Applying conservation of energy and conservation of momentum when appropriate.
- Solving the equations through algebra and substitutions.
- Reflecting on the results and checking for inconsistencies.

Expert problem solvers use this foundation at all levels of physics investigations, from introductory courses through cutting-edge research projects.

11 Equilibrium and Elasticity

Summary

We will explore equilibrium and elasticity in this chapter. We will focus on extended bodies in equilibrium: bodies having no net force or torque acting on them. Bodies deform when forces act on them, so we will examine deformations that describe the stretching, twisting, and compressing of a body. New concepts and principles will be introduced to quantify deformations—ideas based on the concepts and principles we encountered in previous chapters. We will learn about stress, strain, and elastic modulus, and we will further clarify Hooke's law.

Objectives

After studying this chapter, you will understand:

- The conditions required for a body to be in equilibrium.
- The definition of center of gravity and how to apply it to a problem.
- How to solve problems when bodies are in equilibrium.
- How to analyze problems involving the deformation of bodies.
- Stress and strain with respect to tension, compression, and shear forces.
- How to use Young's, bulk, and shear moduli to predict the changes due to stress.
- The limits of stress and strain.

Concepts and Equations

Term	Description
Equilibrium of a Rigid Body	No net force and no net torque acts on a rigid body in equilibrium: $$\sum \vec{F} = 0 \text{ and } \sum \tau = 0.$$
Center of Mass	The torque due to the weight of a body is found by assuming that the entire weight of the body is located at the center of gravity, given by $$\vec{r}_{cm} = \frac{m_1 \vec{r}_1 + m_2 \vec{r}_2 + m_3 \vec{r}_3 + \cdots}{m_1 + m_2 + m_3 + \cdots}$$ The center of gravity is equivalent to the center of mass when gravity is constant.
Stress and Strain	Stress characterizes the strength of a force that stretches, squeezes, or twists an object. Strain is the resulting deformation. Stress and strain are often directly proportional, with the proportionality—the elastic modulus—given by Hooke's law: $$\text{elastic modulus} = \frac{\text{stress}}{\text{stain}}.$$
Tensile and Compressive Stress	Tensile stress is the ratio of the perpendicular component of a force to the cross-sectional area where the force is applied: $$\text{Tensile stress} = \frac{F_\perp}{A}.$$ The SI unit of stress is the pascal (Pa), equal to 1 newton per meter squared. Tensile strain is the ratio of the change in an object's length under stress to its original length: $$\text{Tensile Strain} = \frac{\Delta l}{l_0}.$$ Young's modulus Y is the elastic modulus: $$Y = \frac{\text{Tensile stress}}{\text{Tensile strain}} = \frac{F_\perp / A}{\Delta l / l_0}.$$ Compressive stress and strain are defined in the same manner.
Bulk Stress	The pressure in a fluid is the force per unit area of the fluid: $$p = \frac{F_\perp}{A}.$$ Bulk stress is the change in pressure, and bulk strain is the fractional change in volume, of the fluid. The bulk modulus is the elastic modulus: $$B = \frac{\text{Bulk stress}}{\text{Bulk Strain}} = -\frac{\Delta p}{\Delta V / V_0}.$$

Shear Stress	Shear stress is the force tangent to an object's surface, divided by the area on which the force acts. The shear modulus (S) is the ratio of shear stress to strain: $$S = \frac{\text{Shear stress}}{\text{Shear strain}} = \frac{F_{\parallel}/A}{x/h}.$$
Limits of Hooke's law	There is a maximum stress for which stress and strain are proportional, beyond which Hooke's law is not valid. The elastic limit is the stress beyond which irreversible deformation occurs.

Conceptual Questions

1: Body in equilibrium

A body is acted upon by no net force and no net torque. Is it at rest?

Solution

IDENTIFY, SET UP, AND EXECUTE A body that is moving with constant velocity is in equilibrium. A body could exhibit translational motion with a constant velocity or rotate with a constant angular velocity (or both) and remain in equilibrium.

EVALUATE Just as we saw with forces, constant velocity is a state of equilibrium.

2: Ladder on a frictionless surface

A ladder is placed against a wall. The wall is rough, but the floor is frictionless. Can the ladder be in equilibrium?

Solution

IDENTIFY, SET UP, AND EXECUTE Four forces act on the ladder: the normal force due to the wall, the normal force due to the floor, gravity, and friction due to the wall. Three of the forces act in the vertical direction: gravity (acting downward), friction due to the wall (acting upward), and the normal force due to the floor (acting upward). These forces can sum to zero, since they act in different directions. One force acts in the horizontal direction: the normal force due to the wall. Since there is only one force, the net force on the ladder cannot be zero and therefore the ladder cannot be in equilibrium.

EVALUATE Equilibrium is needed in order for us to use the ladder. How can we achieve equilibrium? Friction with the floor is needed to establish equilibrium. Can the ladder be in equilibrium if placed against a frictionless wall on a rough floor? Yes, the forces and torques can be in equilibrium in this case.

Problems

1: Forces on a diving board

A 4.0-m-long diving board with a uniform mass of 150.0 kg is mounted as shown in Figure 11.1. Find the forces holding the board in place when a 100.0 kg man is standing on the end of the board.

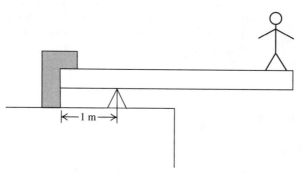

Figure 11.1 Problem 1.

Solution

IDENTIFY We'll use the conditions of equilibrium to solve the problem. The board has two forces holding it in place: a downward force at the left end and an upward force at the pivot. The target variables are the forces acting on the board.

SET UP Figure 11.2 shows the free-body diagram of the board. Forces A and B hold the board in place, the weight of the board acts at the board's center, and the weight of the man acts at the end. We will take counterclockwise torques as positive.

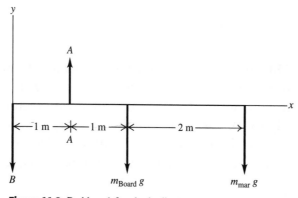

Figure 11.2 Problem 1 free-body diagram.

EXECUTE Newton's first law applied to the board gives

$$\sum F_y = 0 = -F_A + F_B - m_{board}g - m_{man}g.$$

We have two unknowns in this equation, so we need to use the net torque equation. The net torque about the left end is zero:

$$\sum \tau = 0 = F_B(1.0 \text{ m}) - m_{board}g(2.0 \text{ m}) - m_{man}g(4.0 \text{ m}).$$

Solving for the force at B gives

$$F_B = m_{board}g(2.0) + m_{man}g(4.0) = 6860 \text{ N}.$$

Substituting and solving for the force at A gives

$$F_A = F_B - m_{board}g - m_{man}g = 4410 \text{ N}.$$

The force at the left end is 4410 N downward, and the force at the pivot point is 6860 N upward.

EVALUATE To simplify our analysis, we chose the axis for the net torque such that the torque due to force A was zero. We can double check the result by calculating the torque about the pivot point. If we do, we find the same result.

2: Force on a support strut

Find the force exerted by the wall on the uniform strut shown in Figure 11.3 if the strut weighs 100.0 N.

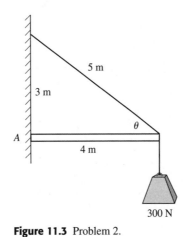

Figure 11.3 Problem 2.

Solution

IDENTIFY We'll use equilibrium conditions to solve the problem. The strut has four forces acting on it; both the net force and the net torque are zero. The target variable is the force acting on the strut due to the wall.

SET UP Figure 11.4 shows the free-body diagram of the strut. The weights of the strut and hanging mass, tension, and the force of the wall act on the strut. We assume that the force of the wall on the strut acts to the right and upward. The forces act in two directions, so we'll need to include net forces in those directions. We will take counterclockwise torques as positive.

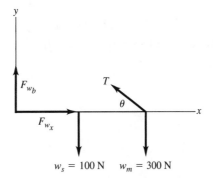

Figure 11.4 Problem 2 free-body diagram.

EXECUTE Newton's first law applied in the x direction gives

$$\sum F_x = 0 = F_{Wx} - T\cos\theta.$$

Newton's first law applied in the y direction gives

$$\sum F_y = 0 = F_{Wy} + T\sin\theta - w_S - w_M.$$

We have three unknowns in these equations, so we need to use the net torque equation. The net torque about the left end of the strut is zero:

$$\sum \tau = 0 = -w_S(2.0\text{ m}) - w_M(4.0\text{ m}) + T\sin\theta(4.0\text{ m}).$$

Inspecting the figure, we see that the sine and cosine of the angle are 3/5 and 4/5, respectively. We solve the torque equation for tension, giving

$$T = \frac{w_S(2.0\text{ m}) + w_M(4.0\text{ m})}{\sin\theta(4.0\text{ m})} = 583\text{ N}.$$

With the tension, we can solve for the components of the force due to the wall:

$$F_{Wx} = T\cos\theta = 467\text{ N},$$
$$F_{Wy} = w_S + w_M - T\sin\theta = 50\,\text{N}.$$

The magnitude of the force is

$$F_W = \sqrt{F_{Wx}^2 + F_{Wy}^2} = 470\text{ N},$$

and it acts at an angle

$$\phi = \tan^{-1}\left(\frac{F_{Wy}}{F_{Wx}}\right) = 83.9°$$

above the positive x-axis.

EVALUATE We see that we chose the correct directions for the force of the wall on the strut. If we hadn't guessed correctly, we would have found negative results for one or both of the force components.

You can double check the result by calculating the torque about any other point. Do you find the same result when you do?

CAUTION **Pick pivots carefully!** Carefully choosing your pivot point simplifies the net torque equation, as we have seen in the previous two examples. Note how the pivot point is chosen in the next two problems.

3: Boom in equilibrium

A horizontal wire supports a boom of length L. The boom supports a 200.0 N weight as shown in Figure 11.5. The boom weighs 200.0 N. Find the tension in the wire and the force exerted by the ground on the boom.

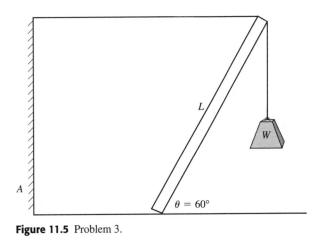

Figure 11.5 Problem 3.

Solution

IDENTIFY We'll use equilibrium conditions to solve the problem. The boom has four forces acting on it; both the net force and the net torque are zero. The target variables are the tension and the force acting on the boom due to the ground.

SET UP Figure 11.6 shows the free-body diagram of the boom. The weights of the boom and the hanging mass, tension, and the force of the floor act on the boom. We assume that the force of the ground on the boom acts to the right and upward. The forces act in two directions, so we'll need to include net forces in those directions. We will take counterclockwise torques as positive.

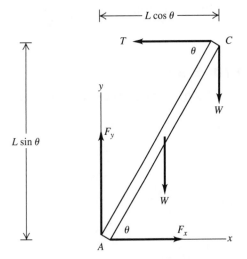

Figure 11.6 Problem 3 free-body diagram.

EXECUTE Newton's first law applied in the x direction gives

$$\sum F_x = 0 = F_x - T.$$

Newton's first law applied in the y direction gives

$$\sum F_y = 0 = F_y - w - w.$$

We have three unknowns in these equations, so we need to use the net torque equation. The net torque about the bottom end of the boom is zero:

$$\sum \tau = 0 = LT\sin\theta - Lw\cos\theta - \tfrac{1}{2}Lw\cos\theta.$$

We solve the torque equation for tension, giving

$$T = \tfrac{3}{2}\frac{w\cos\theta}{\sin\theta} = \tfrac{3}{2}\frac{(200.0\text{ N})\cos 60°}{\sin 60°} = 173\text{ N}.$$

With the tension, we can solve for the components of the force due to the ground:

$$F_x = T = 173\text{ N},$$
$$F_y = -2w = 400.0\text{ N}.$$

The magnitude of the force is

$$F = \sqrt{F_x^2 + F_y^2} = 436\text{ N},$$

and it acts at an angle

$$\phi = \tan^{-1}\left(\frac{F_y}{F_x}\right) = 66.6°$$

above the positive x-axis.

EVALUATE We see that we chose the correct directions for the force of the ground on the boom. If we hadn't guessed correctly, we would have found negative results for one or both of the force components.

You can double check the result by calculating the torque about any other point. Do you find the same result when you do?

4: Coefficient of friction for a strut

Find the minimum coefficient of friction between the weightless horizontal strut and the wall in the system shown in Figure 11.7.

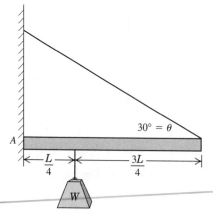

Figure 11.7 Problem 4.

Solution

IDENTIFY We'll use equilibrium conditions to solve the problem. The strut has four forces acting on it; both the net force and the net torque are zero. The target variable is the coefficient of friction at the wall.

SET UP Figure 11.8 shows the free-body diagram of the strut. The weights of the hanging mass, tension, friction, and the normal force of the wall act on the strut. The forces act in two directions, so we'll need to include net forces in those directions. We will take counterclockwise torques as positive.

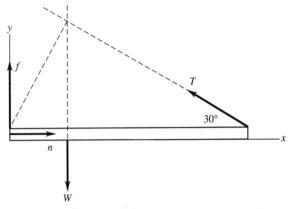

Figure 11.8 Problem 4 free-body diagram.

EXECUTE Newton's first law applied in the x direction gives

$$\sum F_x = 0 = n - T\cos 30°.$$

Newton's first law applied in the y direction gives

$$\sum F_y = 0 = f + T\sin 30° - w.$$

We have three unknowns in these equations, so we need to use the net torque equation. The net torque about the left end of the strut is zero:

$$\sum \tau = 0 = -w\tfrac{1}{4}L + T\sin 30°L.$$

We solve the torque equation for tension, giving

$$T = \frac{w}{4\sin 30°}.$$

With the tension, we can solve for the normal and friction forces:

$$n = T\cos 30° = \frac{w\cos 30°}{4\sin 30°},$$
$$f = w - T\sin 30° = w - \tfrac{1}{4}w = \tfrac{3}{4}w.$$

For there to be no slipping,

$$f < \mu n.$$

Solving for the coefficient of friction gives

$$\mu = \frac{f}{n} = \frac{\tfrac{3}{4}w}{\dfrac{w}{4\tan 30°}} = 3\tan 30° = 1.73.$$

The minimum coefficient of friction is 1.73.

EVALUATE How can you double-check the result? Do you find the same result when you do?

5: Strain in a steel cable

A 10.0 kg weight is hung from a steel wire having an unstretched length of 1.0 m and a diameter of 2.0 mm. How much does the wire stretch?

Solution

IDENTIFY The force acting on the cable is the weight of the mass. Young's modulus will be used to find the change in length.

SET UP We look up Young's modulus for steel and find that it is 2×10^{11} Pa.

EXECUTE Young's modulus is

$$Y = \frac{F/A}{\Delta L/L_0}.$$

Rearranging terms to find the change in length gives

$$\Delta L = \frac{FL_0}{YA} = \frac{mgL_0}{Y\pi(d/2)^2} = \frac{(10.0 \text{ kg})(9.8 \text{ m/s}^2)(1.0 \text{ m})}{(2 \times 10^{11} \text{ N/m}^2)\pi(0.002 \text{ mm}/2)^2} = 1.56 \times 10^{-4} \text{ m} = 0.156 \text{ mm}.$$

EVALUATE We see that the stretch is very small for the wire. This agrees with experience: Steel is a difficult material to stretch.

6: Strain on an elevator cable

A steel elevator cable can support a maximum stress of 9.0×10^7 Pa. If the maximum mass of the fully loaded elevator is 2100 kg and the maximum upward acceleration is 3.0 m/s², what should the diameter of the cable be? By how much does the cable stretch when the elevator is accelerating upward at 3.0 m/s² and 120 m of cable has been released? (Young's modulus for steel is 2×10^{11} Pa.)

Solution

IDENTIFY We'll use Newton's second law to find the tension in the elevator cable and then use the maximum stress to find the cable diameter. To solve the second part, we'll use Young's modulus. The target variables are the cable diameter and the elongation of the cable.

SET UP Figure 11.9 shows the free-body diagram of the elevator. Gravity and tension act on the elevator.

Figure 11.9 Problem 6 free-body diagram.

EXECUTE We'll apply Newton's second law to find the maximum tension in the cable:

$$\sum F_y = T - mg = ma_y.$$

The tension is then

$$T = m(g + a_x) = (2100 \text{ kg})((9.8 \text{ m/s}^2) + (3.0 \text{ m/s}^2)) = 26{,}900 \text{ N}.$$

Stress is the force per unit area, or

$$S = \frac{F}{A} = \frac{T}{A}.$$

The area can be written in terms of the diameter as

$$A = \frac{\pi d^2}{4}.$$

The diameter is then

$$d = 2\sqrt{\frac{T}{S\pi}} = 2\sqrt{\frac{(26{,}900 \text{ N})}{(9.0 \times 10^7)\pi}} = 0.020 \text{ m} = 2.0 \text{ cm},$$

where we have replaced the stress with the maximum stress. Young's modulus then leads to the amount of cable stretch:

$$Y = \frac{l_0 F}{A \Delta l} = \frac{4 l_0 T}{\pi d^2 \Delta l}.$$

Rearranging terms to find the cable stretch gives

$$\Delta l = \frac{4 l_0 T}{\pi d^2 Y} = \frac{4(120 \text{ m})(26{,}900 \text{ N})}{\pi (0.020 \text{ m})^2 (2.0 \times 10^{11} \text{ Pa})} = 0.051 \text{ m} = 5.1 \text{ cm}.$$

The cable must have a 2.0 cm diameter and stretch 5.1 cm when 120 m of cable has been released.

EVALUATE We see that the 2-cm-thick cable stretches over 5 cm. This may appear to be a significant elongation, but it represents only 0.04% of the cable length.

7: Compressibility of oil

Find the compressibility of a 0.1 m³ sample of oil whose volume decreases 2.04×10^{-4} m³ when subjected to an increase in pressure of 1.02×10^7 Pa.

Solution

IDENTIFY AND SET UP We will use the bulk modulus to find the compressibility. With the given information, the compressibility follows directly from the definition of the bulk modulus.

EXECUTE The bulk modulus is given by

$$B = -\frac{\Delta p}{\Delta V / V_0}.$$

Solving, we have

$$B = -\frac{\Delta p V_0}{\Delta V} = -\frac{(1.02 \times 10^7 \text{ Pa})(0.1 \text{ m}^3)}{(-2.04 \times 10^{-4} \text{ m}^3)} = 5.0 \times 10^9 \text{ Pa} = 4.9 \times 10^4 \text{ atm}.$$

The compressibility is the inverse of the bulk modulus, or

$$k = \frac{1}{B} = \frac{1}{4.9 \times 10^4 \text{ atm}} = 2.0 \times 10^{-5}/\text{ atm.}$$

EVALUATE We see that the oil does not compress when subjected to pressure. It requires an increase of 500 atmospheres of pressure to change the volume by 1%.

Try It Yourself!

1: Tension in support cable

Find the tension in the supporting cable and the force acting on the strut due to the wall for the weightless horizontal strut shown in Figure 11.10.

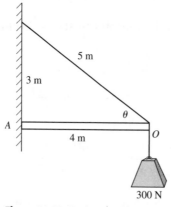

Figure 11.10 Try it yourself 1.

Solution Checkpoints

IDENTIFY AND SET UP The strut is in equilibrium, so apply equilibrium conditions to solve the problem. Start with a free-body diagram, and include the force due to the weight hanging off the end of the strut, tension, and the force of the wall. Set the net forces and net torques equal to zero.

EXECUTE Newton's first law applied in the x and y directions gives

$$\sum F_x = 0 = F_{\text{wall}\,x} - T\cos\theta,$$
$$\sum F_y = 0 = F_{\text{wall}\,y} + T\sin\theta - w.$$

The net torque about the right end of the strut is zero:

$$\sum \tau = 0 = -F_{\text{wall}\,y}(4.0 \text{ m}).$$

Determine the sine and cosine of the angle and solve.

The tension is 500.0 N. The force due to the wall is 400 N, directed perpendicular to the wall.

EVALUATE Choosing a good pivot point simplifies calculations. By calculating the torques about the right end of the strut, we immediately learned that there is no y component of force due to the wall. Do you get the same result if you set the pivot on the left end of the strut?

2: Force acting on a ladder

A ladder of mass 25.0 kg rests against a frictionless wall as shown in Figure 11.11. Find all the forces acting on the ladder.

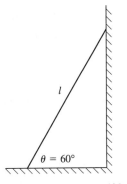

Figure 11.11 Try it yourself 2.

Solution Checkpoints

IDENTIFY AND SET UP The ladder is in equilibrium, so apply equilibrium conditions to solve the problem. Start with a free-body diagram, and include the forces due to the wall, the ground, and gravity. Set the net forces and net torques equal to zero.

EXECUTE Newton's first law applied in the x and y directions gives

$$\sum F_x = 0 = F_{\text{ground}\,x} - n,$$
$$\sum F_y = 0 = F_{\text{ground}\,y} - w.$$

The net torque about the bottom end of the ladder is zero:

$$\sum \tau = 0 = -\frac{L}{2} w \cos\theta + L n \sin\theta.$$

The normal force due to the wall is 71 N. The force due to the ground is 71 N in the positive x direction and 245 N upward.

EVALUATE Does the wall exert a force in the vertical direction? How can you check your results?

3: Friction force acting on a ladder

A ladder of mass 25.0 kg rests against a frictionless wall as shown in Figure 11.11. What is the minimum coefficient of friction between the ladder and the ground that allows the ladder to stand without slipping?

Solution Checkpoints

IDENTIFY AND SET UP The x component of the force due to the ground is the friction force in the previous problem. Use the definition of friction to solve for the coefficient.

EXECUTE The static friction force is

$$F_{\text{ground}\,x} = f \le \mu_s n.$$

The coefficient of static friction is 0.29.

EVALUATE: How did you find the normal force?

4: Stress in wire

A copper wire of cross-sectional area 0.050 cm² and length 5.0 m is attached end to end to a steel wire of length 3.0 m and cross-sectional area 0.020 cm². The wires are stretched under a tension of 200.0 N. Find the stress in each wire and the total change in length for the combination.

Solution Checkpoints

IDENTIFY Tensile stress is the force per area. Young's modulus will be used to find the change in length. You can find Young's modulus for steel and copper from Table 11.1 of the text. Then add the changes in lengths to find the total change in length.

EXECUTE The stresses are

$$\text{Stress}_{\text{Copper}} = \frac{F}{A} = 4.0 \times 10^7 \text{ Pa},$$

$$\text{Stress}_{\text{Steel}} = \frac{F}{A} = 1.0 \times 10^8 \text{ Pa}.$$

The change in length of the copper is

$$\Delta L = \frac{FL_0}{YA} = 1.5 \times 10^{-3} \text{ m}.$$

For steel, the change in length is 1.8×10^{-3} m. The total length is 3.3×10^{-3} m, or 3.3 mm.

EVALUATE We see that the stretch is very small for combined length of wire.

12 Gravitation

Summary

In this chapter, we will delve into the gravitational interaction by learning that the gravity we experience on earth also applies to planets and celestial objects and is responsible for their motion. We will see how to apply Newton's law of gravitation and gain a better understanding of the concept of weight. We'll use this knowledge to explain the orbits of satellites and planets. We will also examine an extreme case of gravity: black holes.

Objectives

After studying this chapter, you will understand

- How to apply Newton's law of gravitation to pairs of masses.
- The general definition of weight.
- How to use the generalized expression for gravitational potential energy.
- How satellites orbit astronomical bodies.
- How to predict the motion of satellites.
- Kepler's three laws of planetary motion.
- The definition of a black hole and the properties of black holes.

Concepts and Equations

Term	Description
Newton's Law of Gravitation	Newton's law of gravitation states that the magnitude of the force between two bodies with masses m_1 and m_2, separated by a distance r, is given by $$F_g = G\frac{m_1 m_2}{r^2},$$ where G denotes the gravitational constant and is equal to $6.67 \times 10^{-11}\ \text{N} \cdot \text{m}^2/\text{kg}^2$. The gravitational force is always attractive and is directed along the line that separates the objects.
Weight	The weight of an object is the total gravitational force exerted on the object by all other objects in the universe. Near the surface of the earth, an object's weight is very nearly equal to the gravitational force of the earth on the object alone.
Gravitational Potential Energy	The gravitational potential energy of two bodies with masses m and m_E, separated by a distance r, is given by $$U = -\frac{Gm_E m}{r}.$$ The potential energy is never positive and is zero only when the two objects are infinitely far apart.
Orbits of Satellites	For a satellite moving in a circular orbit, the gravitational attraction between the satellite and the astronomical body provides the centripetal acceleration. The velocity v and period T of a satellite orbiting at a radius r are given, respectively, by $$v = \sqrt{\frac{GM}{r}},$$ $$T = \frac{2\pi r^{3/2}}{\sqrt{GM}},$$ where M is the mass of the astronomical body.
Kepler's Laws	Kepler's three laws describe the motion of a planet or satellite around the sun or another planet. They describe the elliptical motion and the area swept out per unit time in the orbit, as well as relate the period of the planet or satellite to the major axis of its orbit.
Black Holes	A black hole is a nonrotating spherical mass distribution with total mass M contained within a radius R_S, the Schwarzschild radius, given by $$R_S = \frac{2GM}{c^2}.$$ Gravity prevents matter and light from escaping from within a sphere with radius R_S.

Conceptual Questions

1: Is the earth falling?

There is a net gravitational force between the earth and the sun, so why doesn't the earth fall into the sun?

Solution

IDENTIFY, SET UP, AND EXECUTE The earth is constantly falling toward the sun, but the earth doesn't get closer to the sun, since the sun's surface curves away beneath the earth. Recall projectile motion from Chapter 3. If we launch an object parallel to the ground, it follows a parabolic path. If we give the object a larger initial velocity, then the object moves farther away from the launch site as it falls. The earth is round, so as the object moves farther away, the object will have a larger distance to fall. At a sufficiently high launch velocity, the object will make a complete revolution and not land on the ground. This is the same physical situation as the earth revolving around the sun. If the earth had a lower velocity, it would fall into the sun.

EVALUATE This result may seem a bit odd, but is indeed accurate. The result also shows how our understanding of one physical phenomenon helps us understand other phenomena: Our experience with projectile motion helped us interpret the motion of the earth around the sun.

2: Does the earth maintain a constant speed?

In the previous question, we saw that the earth is constantly falling. Does it maintain a constant speed?

Solution

IDENTIFY, SET UP, AND EXECUTE There is a net gravitational force acting on the earth due to the sun. The direction of the net force is toward the sun; however, the earth's velocity is perpendicular to the direction of force. The gravitational force can change only the direction of the earth's velocity around the sun and not the magnitude of the velocity. Thus, the earth maintains a constant speed as it orbits the sun. The earth does *not* maintain a constant velocity, since its direction is always changing.

EVALUATE We've come to know that a net force causes acceleration—a change in velocity. In the previous chapters, often the magnitude of an object's velocity changed when the object was acted upon by a net force. This chapter examines additional consequences of the influences of forces.

3: Two satellites and the earth

The moon and the international space station are located on opposite sides of the earth. How does the presence of the earth influence the gravitational force between the moon and the space station?

Solution

IDENTIFY, SET UP, AND EXECUTE The gravitational force between two bodies depends only on the mass of the bodies and their separation, according to Newton's law of gravitation. The force between the moon and the space station is not affected by the presence of the earth. There are forces between the earth and the two orbiting satellites, but those forces do not affect the force between the satellites.

EVALUATE Forces are between two bodies. The net force on a single body may include forces due to many bodies, but each force acts between two bodies.

Problems

1: Gravitational force due to three masses

Three masses are arranged as shown in Figure 12.1. Find the net force acting on the top mass (*A*). Each mass is 5.00 kg.

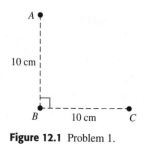

Figure 12.1 Problem 1.

Solution

IDENTIFY The net force on *A* is found by adding the force on *A* due to *B* and the force on *A* due to *C*. The target variable is the force on *A*.

SETUP A free-body diagram illustrating the two forces on *A* is shown in Figure 12.2. We'll need to add the two forces by using components. Newton's law of gravitation gives the magnitude of the forces. We'll use the coordinate axes provided in the figure.

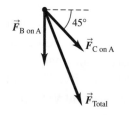

Figure 12.2 Problem 1 free-body diagram.

EXECUTE To apply Newton's law of gravity, we need the distances between the masses. The distance between *A* and *B* is 10.0 cm. By summing the squares of the sides of the triangle and taking the square root, we find that the distance between *A* and *C* is 14.1 cm. The force of *B* on *A* is

$$F_{B\,on\,A} = \frac{Gm_B m_A}{r_{BA}^2} = \frac{(6.67 \times 10^{-11}\,\text{N} \cdot \text{m}^2/\text{kg}^2)(5.00\,\text{kg})(5.00\,\text{kg})}{(0.100\,\text{cm})^2} = 1.67 \times 10^{-7}\,\text{N}.$$

The force of *C* on *A* is

$$F_{C\,on\,A} = \frac{Gm_C m_A}{r_{CA}^2} = \frac{(6.67 \times 10^{-11}\,\text{N} \cdot \text{m}^2/\text{kg}^2)(5.00\,\text{kg})(5.00\,\text{kg})}{(0.141\,\text{cm})^2} = 8.39 \times 10^{-8}\,\text{N}.$$

With the magnitudes of the forces determined, we simply add the two vectors together, using components. The force of *C* on *A* has the only *x* component:

$$\sum F_x = F_{C\,on\,A} \sin 45° = 5.93 \times 10^{-8}\,\text{N}.$$

The 45° angle results from the masses arranged as an isosceles triangle. Both forces have y components:

$$\sum F_y = -F_{B \text{ on } A} - F_{C \text{ on } A} \sin 45° = -2.26 \times 10^{-7} \text{ N}.$$

The negative result indicates that the y component points downward. We find the magnitude of the net force by combining the components:

$$F = \sqrt{F_x^2 + F_y^2} = 2.34 \times 10^{-7} \text{ N}.$$

The direction of the net force is found by using the tangent. We want to specify the angle ϕ with respect to the x-axis:

$$\phi = \tan^{-1}\frac{F_y}{F_x} = \tan^{-1}\frac{(-2.26 \times 10^{-7} \text{ N})}{(5.93 \times 10^{-8} \text{ N})} = -75.3°.$$

The net force on A has magnitude 2.34×10^{-7} N and points 75.3° below the positive x-axis.

EVALUATE We see that the gravitational force between the masses is very small. To have an appreciable gravitational force, we need at least one large mass, such as the earth. Also, we can see that Newton's third law is valid: Reversing indices in the first two equations would result in a force of the same magnitude, but opposite in direction.

2: Orbit of a weather satellite

Imagine you are designing a new weather satellite. The goal is to have the satellite orbit the earth in a circular orbit every 6 hours. At what distance above the earth's surface should the satellite be placed to obtain the correct period?

Solution

IDENTIFY The force acting on the satellite is the force of gravitation between the satellite and the earth. The satellite follows a circular orbit and so has a radial acceleration toward the center of the earth. The target variable is the height of the satellite's orbit.

SET UP Newton's law of gravitation gives the force on the satellite due to the earth. The acceleration of the satellite is centripetal. Combining the two equations will lead to the velocity of the satellite and the period of rotation. We solve for the distance by setting the period to 6 hours.

EXECUTE Newton's law of gravitation gives the force on the satellite, namely,

$$F_g = \frac{Gmm_E}{r^2},$$

where m is the mass of the satellite, m_E is the mass of the earth, and r is the distance from the center of the earth to the satellite. Newton's second law gives the net force on the satellite (the acceleration is v^2/r):

$$\sum F = ma_{\text{rad}}, \sum F = F_g = \frac{Gmm_E}{r^2} = m\frac{v^2}{r}.$$

Solving for v, we find that

$$v = \sqrt{\frac{Gm_E}{r}}.$$

We can also write the velocity in terms of the distance the satellite travels $(2\pi r)$ in one period (T):

$$v = \frac{2\pi r}{T}.$$

To find the radius, we equate the last two equations and solve for r, obtaining

$$r = \sqrt[3]{\frac{Gm_E T^2}{4\pi^2}} = \sqrt[3]{\frac{(6.67 \times 10^{-11}\,\text{N} \cdot \text{m}^2/\text{kg}^2)(5.98 \times 10^{24}\,\text{kg})(2.16 \times 10^4\,\text{s})^2}{4\pi^2}} = 1.68 \times 10^7\,\text{m},$$

where we replaced the 6 hour period with the equivalent 21,600 s. The satellite should be placed in an orbit of radius 16,800 km. Subtracting the radius of the earth (6380 km) from the radius of the satellite's orbit, we find that the satellite should be placed 10,400 km above the earth's surface to achieve a 6 hour orbital period.

EVALUATE We could have avoided our derivation and used the textbook's equation 12.12 to arrive at the solution directly. However, this review helps remind us how to find the solution without searching the book.

Practice Problem: What does the free-body diagram of the satellite look like? *Answer:* A single vector.

3: Velocity of rocket

What velocity must a rocket have at the surface of the earth if it is to rise to a height equal to the earth's radius before it begins to descend? Ignore air resistance.

Solution

IDENTIFY We'll use energy conservation to solve the problem. The target variable is the initial velocity of the rocket.

SET UP The rocket is given an initial kinetic energy for lift-off and has initial gravitational potential energy. At the top of the flight, its kinetic energy drops to zero, leaving only gravitational potential energy. We'll set these equal to each other to solve for the initial velocity.

EXECUTE Energy conservation gives

$$K_1 + U_1 = K_2 + U_2.$$

Replacing both sides with the expressions for the two forms of energy gives

$$\frac{1}{2}mv^2 - \frac{Gmm_E}{r_E} = -\frac{Gmm_E}{2r_E},$$

where r_E is the radius of the earth and $K_2 = 0$. Solving for the initial velocity, we obtain

$$v = \sqrt{\frac{Gm_E}{r_E}} = \sqrt{\frac{r_{\text{earth}}^2 g}{r_E}} = \sqrt{g r_E} = \sqrt{(9.8\,\text{m/s}^2)(6.36 \times 10^6\,\text{m})} = 7.9 \times 10^3\,\text{m/s}.$$

The initial velocity of the rocket must be $7.9 \times 10^3\,\text{m/s}$.

EVALUATE By examining the solution to the problem, we see why rockets must have large initial velocities to be propelled into space. The actual value is higher due to air resistance.

4: What if the sun were a black hole?

What would the sun's radius need to be in order for the surface escape velocity to be *c?*

Solution

IDENTIFY AND SET UP A radius corresponding to an escape velocity of *c* is the Schwarzschild radius. We can use the expression for the Schwarzschild radius to solve the problem.

EXECUTE The Schwarzschild radius is given by

$$R_S = \frac{2GM}{c^2}.$$

Substituting gives

$$R_S = \frac{2GM}{c^2} = \frac{2(6.67 \times 10^{-11} \,\text{Nm}^2/\text{kg}^2)(1.99 \times 10^{30} \,\text{kg})}{(3.0 \times 10^8 \,\text{m/s})^2} = 2960 \,\text{m}.$$

The radius would be 2,960 m.

EVALUATE The sun would have to be compressed by a factor of over 200,000 to become a black hole.

Practice Problem: What is the Schwarzschild radius for the earth? *Answer*: 8.8 mm.

Try It Yourself!

1: Sun's gravity on earth

The sun is a distance of 1.48×10^{11} m from earth and has a mass of 1.99×10^{30} kg. Find the ratio of the sun's gravitational force to the earth's gravitational force on an object on the earth's surface.

Solution Checkpoints

IDENTIFY AND SET UP Newton's law of gravitation is used to find the force of gravity due to the earth and the force of gravity due to the sun. Taking their ratio solves the problem.

Substituting the values given results in a ratio of 6.03×10^{-4} m/s.

EVALUATE Do you need to include the force of gravity due to the sun in physics problems on earth?

2: Three masses positioned in a triangle

Three 1000.0 kg masses are at the vertices of an equilateral triangle with sides of length 1.0 m. Find the force due to any two masses on the third.

Solution Checkpoints

IDENTIFY AND SET UP Newton's law of gravitation is used to find the force of gravitation due to the other masses. The two forces must be added as vectors. A free-body diagram should be used.

EXECUTE The force between two masses is

$$F_g = \frac{Gm_1m_2}{r^2} = 6.67 \times 10^{-5} \,\text{N}.$$

The net force is 1.16×10^{-4} N, directed toward the line separating the other two masses.

EVALUATE Did you use symmetry arguments to determine that the force has no component parallel to the line separating the two masses?

3: Orbit of a communications satellite

Communications satellites revolve in orbits over the earth's equator, adjusted so that their period of rotation is the same as the period of rotation of the earth about its axis. This speed and a period of rotation together cause the satellite to remain in a fixed position in the sky. Find the height of these satellites' orbit above the earth.

Solution Checkpoints

IDENTIFY AND SET UP You can find the height by setting the period equal to 24 hours. The solution can be found by solving Newton's second law or by using the equation in the book.

EXECUTE The period is given by

$$T = \frac{2\pi r^{3/2}}{\sqrt{gR^2}}.$$

The terms in this equation can be rearranged to solve for r. The height above earth is r minus the radius of earth. The satellite must be placed 36,000 km above the surface of the earth to remain in geosynchronous orbit.

EVALUATE Such heights require additional power for radio signals to reach the satellites and for the signal to be redirected back to earth.

4: Escape from the sun

What is the escape velocity of a particle on the surface of the sun?

Solution Checkpoints

IDENTIFY AND SET UP The escape velocity can be found from energy conservation or the equation given in the text. Parameters for the sun can be found in the chapter or appendix.

EXECUTE The escape velocity is given by

$$v = \sqrt{\frac{2GM}{r}}.$$

After substituting, we find the escape velocity to be 6.48×10^5 m/s.

EVALUATE Is the resulting velocity greater or less than the escape velocity of a particle on the surface of the earth? Why?

13 Periodic Motion

Summary

We will examine periodic motion, or oscillation, in this chapter. Many systems exhibit periodic motion, such as a swinging pendulum, a ball on a spring, or the membrane of a drum. We will describe the motion of oscillating bodies, characterized by amplitude, period, frequency, and angular frequency. We'll use force equations and energy concepts to analyze their motions. We will look at several models of periodic motion that can be used to represent the motion of many oscillators. Periodic motion plays a vital role in many areas of physics, and this chapter will lay the foundation for further studies.

Objectives

After studying this chapter, you will understand

- Periodic motion and the terminology used to describe oscillations.
- How to identify and analyze simple harmonic motion.
- Energy and motion as a function of time for a particle in simple harmonic motion.
- The simple pendulum, the physical pendulum, damped and forced oscillations, and resonance.

Concepts and Equations

Term	Description
Periodic Motion	Periodic motion is motion that repeats in a definite cycle. Periodic motion occurs when an object is displaced from its equilibrium position and a restoring force exists that tends to return the object to equilibrium. The amplitude is the maximum magnitude of displacement from equilibrium. A cycle is one complete round-trip. The period is the time taken to complete one cycle. Frequency (f) is the number of cycles per unit time. Angular frequency (ω) is 2π times the frequency. Period, frequency, and angular frequency are related: $$T = \frac{1}{f}, \qquad f = \frac{1}{T}, \qquad \omega = 2\pi f = \frac{2\pi}{T}.$$
Simple Harmonic Motion	Simple harmonic motion (SHM) is periodic motion in which the restoring force is directly proportional to the object's displacement. Often, SHM occurs when the displacement is small. The equation of motion is $$x = A\cos(\omega t + \phi).$$ A system attached to a spring with spring constant k and having mass m will oscillate at a frequency of $$\omega = 2\pi f \frac{\omega}{2\pi} = \sqrt{\frac{k}{m}}$$ and a period of $$T = \frac{1}{f} = 2\pi\sqrt{\frac{m}{k}}.$$ The total mechanical energy remains constant in SHM and can be expressed in terms of its amplitude: $$E = \tfrac{1}{2}mv^2 + \tfrac{1}{2}kx^2 = \tfrac{1}{2}kA^2.$$ For angular simple harmonic motion, the frequency is related to the moment of inertia and the torsion constant by $$\omega = \sqrt{\frac{\kappa}{I}}.$$
Simple Pendulum	A simple pendulum is a model of a point mass suspended by a massless string in a gravitational field. For small displacements, a pendulum of length L has frequency $$\omega = 2\pi f = \sqrt{\frac{g}{L}}.$$
Physical Pendulum	A physical pendulum is any body suspended from an axis of rotation. The angular frequency for small-amplitude oscillations is given by $$\omega = \sqrt{\frac{mgd}{I}}.$$
Damped Oscillations	A simple harmonic oscillator impelled by a force that is proportional to velocity exhibits *damped oscillations*. The angular frequency becomes $$\omega' = \sqrt{\frac{k}{m} - \frac{b^2}{4m^2}}.$$

Systems are called critically damped, overdamped, and underdamped according to how they return to equilibrium.

Forced Oscillations	Periodic motion in a system with a sinusoidally varying driving force is called *forced oscillation* or *driven oscillation*. Resonance occurs when the driving angular frequency is near the natural-oscillation angular frequency, increasing the amplitude of the motion. The amplitude is given by $$A = \frac{F_{max}}{\sqrt{(k - m\omega_d^2)^2 + b\omega_d^2}}.$$

Conceptual Questions

1: Glider in simple harmonic motion

A glider attached to a spring and set on a horizontal air track is allowed to oscillate with a 5.0 cm amplitude. How far does the glider travel in one period?

Solution

IDENTIFY, SET UP, AND EXECUTE We answer the question by considering how the glider moves during one period. The period is the time an object takes to move from any position through one complete periodic cycle and return to the starting position. Imagine that the glider starts from an equilibrium position, moves to the right, and momentarily stops at a displacement equal to the amplitude. It has traveled a distance of one amplitude, or 5.0 cm. The glider then returns to its equilibrium position, traveling a second distance equal to the amplitude, or a total of 10.0 cm. The glider continues moving to the left until it reaches its maximum displacement on the left side, thus traveling a third distance equal to the amplitude (15.0 cm total). The glider then moves to the right and returns to the starting position, traveling a fourth distance equal to the amplitude (20.0 cm total).

In one period, the glider travels a distance equal to four amplitudes, or 20.0 cm.

EVALUATE You must distinguish between amplitude and total distance traveled. Comprehending this difference helps build an understanding of simple harmonic motion.

2: Gravity on the moon

You are asked to estimate the moon's gravitational acceleration by watching a video of the early lunar explorations. How could you estimate the acceleration due to gravity on the moon?

Solution

IDENTIFY, SET UP, AND EXECUTE We've seen that, for small oscillations, the period of a simple pendulum is related to the gravitational constant and the length of the pendulum. If you can find an object that can be approximated by a simple pendulum, then you can determine the gravitational acceleration from the object's motion. One approach would be to look for a dangling object during a moonwalk. You can estimate the length of the pendulum by comparing it with the size of the astronaut on the walk and measure the time with a stopwatch or by counting video frames.

EVALUATE The moon's gravitational acceleration was estimated by physics students around the world watching the early moonwalks. The technique can also be used to estimate the sizes of objects in videos by taking the known gravitational acceleration value and combining it with the period to find the length of the pendulum.

Problems

1: Mass on a spring

A spring stretches 4.7 cm from its equilibrium position when a 1.2 kg mass is hung from it. If the mass is now stretched 6.5 cm from the equilibrium position and released, find (a) the period of the motion, (b) the maximum velocity of the mass, and (c) the maximum acceleration of the mass.

Solution

IDENTIFY Since the net force acting on the block is proportional to the displacement of the block, the motion is simple harmonic motion. The target variables are the period, maximum velocity, and maximum acceleration.

SET UP We'll use the equations of simple harmonic motion to find the solutions to the problem. We first find the spring constant, using the preliminary information.

EXECUTE We find the spring constant from Hooke's law. When the mass is initially attached to the spring, it hangs in equilibrium, so the spring force is equal to the product of the mass and the acceleration due to gravity:

$$F_s = kx = mg.$$

The spring constant is

$$k = \frac{mg}{x} = \frac{(1.2 \text{ kg})(9.8 \text{ m/s}^2)}{0.047 \text{ m}} = 250 \text{ N/m}.$$

With the spring constant, we can directly find the period:

$$T = 2\pi\sqrt{\frac{m}{k}} = 2\pi\sqrt{\frac{(1.2 \text{ kg})}{(250 \text{ N/m})}} = 0.44 \text{ s}.$$

The maximum velocity is

$$v_{max} = \sqrt{\frac{k}{m}}A = \sqrt{\frac{(250 \text{ N/m})}{(1.2 \text{ kg})}}(0.065 \text{ m}) = 0.94 \text{ m/s}.$$

The maximum (positive) acceleration occurs when the mass is at its most negative position, so

$$a_{max} = -\frac{k}{m}x = -\frac{(250 \text{ N/m})}{(1.2 \text{ kg})}(-0.065 \text{ m}) = 13.5 \text{ m/s}^2.$$

The mass oscillates with a period of 0.44 s and has a maximum velocity of 0.94 m/s and a maximum acceleration of 13.5 m/s², upward.

EVALUATE Simple harmonic motion is the most complicated motion we have studied to date. However, our previous experiences led to straightforward relationships from which we can easily extract useful information.

We could also have found the motion as a function of time, taken derivatives to find the velocity and acceleration as a function of time, and then determined the maxima—the amplitudes of the velocity and acceleration functions.

2: Object in SHM

A 200.0 g mass vibrates in SHM with a total energy of 25.0 J and a frequency of 5.0 Hz. Find the time it takes to move from 25.0 cm below to 25.0 cm above the equilibrium position.

Solution

IDENTIFY We will use the simple harmonic motion relations to find the solution. The target variable is the time needed to move the specified distance.

SET UP We'll use the equation of simple harmonic motion to find the solutions to the problem. We will find the times the mass is -0.25 cm and $+25.0$ cm from the equilibrium position. We will need to find the spring constant and amplitude from the information given.

EXECUTE The spring constant can be found from the frequency equation:

$$f = \frac{1}{2\pi}\sqrt{\frac{k}{m}}.$$

Solving for k gives

$$k = m(2\pi f)^2 = (0.200 \text{ kg})(2\pi(5.0 \text{ Hz}))^2 = 197.4 \text{ N/m}.$$

The total energy is given by

$$E = \tfrac{1}{2}mv^2 + \tfrac{1}{2}kx^2.$$

The amplitude is the maximum displacement. At the maximum displacement, the velocity is zero. Solving for the amplitude results in

$$E = \tfrac{1}{2}kA^2,$$

$$A = \sqrt{2E/k} = \sqrt{2(24.7 \text{ J})/(197 \text{ N/m})} = 0.500 \text{ m}.$$

The position as a function of time is given by

$$x = A\sin\omega t = A\sin 2\pi ft.$$

We need to solve for the time when $x = -0.25$ m and $x = 0.25$ m. Solving gives

$$t_{-0.25 \text{ m}} = \frac{1}{2\pi f}\sin^{-1}\left(\frac{x}{A}\right) = \frac{1}{2\pi(5.0 \text{ Hz})}\sin^{-1}\left(\frac{-0.25 \text{ m}}{0.50 \text{ m}}\right) = -0.0167 \text{ s},$$

$$t_{+0.25 \text{ m}} = \frac{1}{2\pi f}\sin^{-1}\left(\frac{x}{A}\right) = \frac{1}{2\pi(5.0 \text{ Hz})}\sin^{-1}\left(\frac{0.25 \text{ m}}{0.50 \text{ m}}\right) = 0.0167 \text{ s}.$$

The time the mass takes to move from 25.0 cm below to 25.0 cm above the equilibrium position is 0.0333 s.

EVALUATE We see that the time the mass takes to move from half the amplitude below to half the amplitude above the equilibrium position is about 15% of the period. Does this make sense? Yes, it makes sense, since the velocity near the equilibrium point is maximal.

Practice Problem: How long would it take the mass to move from the equilibrium point to the maximum amplitude? *Answer*: 0.05 s, or one-fourth the period.

CAUTION **Use radians!** When you work with trigonometric functions, the arguments are in radians. You must either set your calculator to radian mode or convert degrees to radians after taking the inverse of the trigonometric function.

3: Period of a simple pendulum

A simple pendulum reaches a maximum angle of 7.2° after swinging through the bottom of its path with a maximum speed of 0.35 m/s. What is the period of the pendulum's oscillation?

Solution

IDENTIFY The period of a simple pendulum depends on its length and the gravitational constant. We can find the target variable—the length—from the velocity and maximum angle.

SET UP The maximum angle, arc length, and length are related to each other. The amplitude and the maximum speed will be used to find the length of the pendulum, and the period will be derived from the length.

EXECUTE The amplitude of a simple pendulum is the maximum arc length, which is related to the maximum angle by the length:

$$S_{max} = L\theta_{max}.$$

The maximum velocity is

$$v_{max} = 2\pi fA = 2\pi fL\theta_{max}.$$

For a simple pendulum, the frequency is found from the length:

$$f = \frac{1}{2\pi}\sqrt{\frac{g}{L}}.$$

Combining these equations, we obtain the length:

$$v_{max} = 2\pi\left(\frac{1}{2\pi}\sqrt{\frac{g}{L}}\right)L\theta_{max} = \sqrt{gL}\theta_{max},$$

$$L = \frac{v_{max}^2}{g\theta_{max}^2} = \frac{(0.35 \text{ m/s})^2}{(9.8 \text{ m/s}^2)(0.126)^2} = 0.79 \text{ m}.$$

Note that we replaced the maximum angle of 7.2° with the equivalent 0.126 radian. The period is then

$$T = 2\pi\sqrt{\frac{L}{g}} = 2\pi\sqrt{\frac{(0.79 \text{ m})}{(9.8 \text{ m/s}^2)}} = 1.8 \text{ s}.$$

The pendulum has a length of 79 cm and a period of 1.8 s.

EVALUATE The period of a simple pendulum depends only on the length of the pendulum and the gravitational constant. The maximum angle and velocity provided enough information to solve the problem.

4: Period of a physical pendulum

A thin, uniform rod is pivoted at a point one-quarter of its length from one end and is then pivoted at a point at its end. Find the ratio of the two periods.

Solution

IDENTIFY The period of a physical pendulum depends on its moment of inertia, its mass, the distance to its center of mass, and the gravitational constant. The target variable is the ratio of the periods for the two pivot points.

SET UP We will calculate the moment of inertia for each of the two pivot points and then combine the two moments to form the ratio.

EXECUTE The moment of inertia of a rod about its end point is

$$I_{end} = \frac{1}{3}ML^2.$$

When the rod is pivoted at a point one-quarter along its length, the moment of inertia is found by the parallel-axis theorem:

$$I_{1/4} = I_{end} + mx^2 = \frac{1}{3}ML^2 + M\left(\frac{L}{4}\right)^2 = \frac{7}{48}ML^2.$$

The period of a physical pendulum is given by

$$T = 2\pi\sqrt{\frac{I}{mgd}}.$$

The ratio of the two periods is then

$$\frac{T_{1/4}}{T_{end}} = \frac{2\pi\sqrt{\dfrac{I_{1/4}}{mgd_{1/4}}}}{2\pi\sqrt{\dfrac{I_{end}}{mgd_{end}}}} = \sqrt{\frac{I_{1/4}d_{end}}{I_{end}d_{1/4}}} = \sqrt{\frac{\dfrac{7}{48}ML^2\dfrac{L}{2}}{\dfrac{1}{3}ML^2\dfrac{L}{4}}} = \sqrt{\frac{(7)(3)(4)}{(2)(48)}} = 0.94.$$

The ratio of the period when the rod is pivoted at a point one-quarter of its length from one end to the period when the rod is pivoted at a point at its end is 0.94.

EVALUATE We see that the period does not change substantially when the pivot point moves between the two positions.

5: Oscillating blocks

Two blocks shown in Figure 13.1 oscillate on a frictionless surface with a frequency of 0.30 Hz. The top block has a mass of 2.0 kg and the bottom block has a mass of 4.5 kg. If the amplitude is increased to 25 cm, the top block begins to slide. What is the coefficient of static friction?

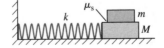

Figure 13.1 Problem 5.

Solution

IDENTIFY When the top block just begins to slide, the force applied must be equal to the maximum static frictional force. The maximum applied force occurs at the maximum displacement (equal to the amplitude). We will determine the force at the turning point and use that to solve for the coefficient of friction, the target variable.

SET UP We'll need the spring constant, which we extract from the initial frequency. The motion is simple harmonic motion, as the only horizontal force acting on the blocks is the spring force, a restoring force that is directly proportional to the displacement.

EXECUTE The frequency in simple harmonic motion depends on the spring constant according to the formula

$$f = \frac{1}{2\pi}\sqrt{\frac{k}{m + M}},$$

where we include the combined mass of the oscillating blocks. Solving for k gives

$$k = (2\pi f)^2(m + M) = (2\pi(0.30\text{ Hz}))^2((2.0\text{ kg}) + (4.5\text{ kg})) = 23.1\text{ N/m}.$$

Recall that the maximum static frictional force is $f_s = \mu_s n$. For the top block, the normal force is mg. The maximum force applied by the spring is kA. Equating these forces gives

$$f_s = \mu_s n = \mu_s mg = kA.$$

Rearranging terms to find the coefficient of static friction yields

$$\mu_s = \frac{kA}{mg} = \frac{(23.1\text{ N/m})(0.25\text{ m})}{(2.0\text{ kg})(9.8\text{ m/s}^2)} = 0.29.$$

The coefficient of static friction between the blocks is 0.29.

EVALUATE This problem brings together topics from several areas we have studied throughout the text, including the normal force, frictional forces, the spring force, and simple harmonic motion. Combining our knowledge helps us understand complex phenomena.

6: Damped oscillation

A body with mass 0.30 kg hangs by a spring with force constant 50.0 N/m. By what factor is the frequency of oscillation reduced if the oscillation is damped and reaches $1/e$ of its original amplitude in 100 oscillations?

Solution

IDENTIFY The amplitude in damped SHM diminishes by a factor of $e^{-bt/2m}$ as a function of time. We'll set this quantity to $1/e$ to solve. The target variable is the fractional frequency shift, which is the change in frequency divided by the undamped frequency.

SET UP We will write the fractional frequency shift in terms of the damped and undamped frequencies. We will need to expand the frequency equation to simplify the shift and substitute the exponential decay information to solve.

EXECUTE The undamped frequency is given by

$$\omega = \sqrt{\frac{k}{m}}.$$

The damped frequency is given by

$$\omega' = \sqrt{\frac{k}{m} - \frac{b^2}{4m^2}}.$$

Combining these two equations to find the fractional frequency shift produces

$$\frac{\Delta\omega}{\omega} = \frac{\omega' - \omega}{\omega} = \frac{\sqrt{\dfrac{k}{m} - \dfrac{b^2}{4m^2}} - \sqrt{\dfrac{k}{m}}}{\sqrt{\dfrac{k}{m}}}$$

$$= \sqrt{\left[1 - \left(\frac{m}{k}\right)\frac{b^2}{4m^2}\right]} - 1.$$

For small damping, the second term in the square root is small, so we can simplify by using the approximation

$$\sqrt{1 - x} \approx 1 - \frac{1}{2}x.$$

This gives

$$\frac{\Delta\omega}{\omega} \approx 1 - \frac{1}{2}\left(\frac{m}{k}\right)\frac{b^2}{4m^2} - 1 = \frac{1}{2}\left(\frac{m}{k}\right)\frac{b^2}{4m^2}.$$

We need to eliminate the damping term. We use the exponential decay information. The amplitude in damped oscillations changes as

$$A(t) = A_0 e^{-(b/2m)t}.$$

We know that after 100 oscillations the amplitude drops to $1/e$. This gives

$$e^{-(b/2m)t} = e^{-1}.$$

Taking the logarithm of each side yields an expression for $b/2m$:

$$-(b/2m)[100T] = -1$$

$$\frac{b}{2m} = \frac{1}{100}\frac{1}{T} = \frac{1}{100}\frac{1}{2\pi}\sqrt{\frac{k}{m}}.$$

We now solve for the shift:

$$\frac{\Delta\omega}{\omega} = \frac{1}{2}\left(\frac{m}{k}\right)\frac{b^2}{4m^2} = \frac{1}{2}\left(\frac{m}{k}\right)\left(\frac{1}{100}\frac{1}{2\pi}\sqrt{\frac{k}{m}}\right)^2 = \frac{1}{2}\left(\frac{1}{100}\frac{1}{2\pi}\right)^2 = 1.27 \times 10^{-6}.$$

The frequency shift is 1.27×10^{-6}.

EVALUATE We see that the frequency shift for this problem is very small.

Try It Yourself!

1: SHM practice

A body of mass 0.5 kg is attached to a spring with spring constant 100.0 N/m and is allowed to oscillate on a horizontal frictionless surface. It is given an initial velocity, at $x = 0$, of 5.0 m/s. Find (a) the total energy of the body, (b) the amplitude of oscillation, (c) the velocity when the displacement is half of the amplitude, (d) the displacement when the velocity is half of its initial value, (e) the displacement when the kinetic and potential energies are equal, and (f) the frequency and period of the motion.

Solution Checkpoints

IDENTIFY AND SET UP The body is in simple harmonic motion. Use the energy equations for SHM to solve for the many target variables. Remember that energy is conserved in SHM.

EXECUTE Energy conservation is used to solve (a) through (e) by substituting the appropriate knowns to find the unknowns. Energy conservation for the system states that

$$E = \tfrac{1}{2}mv^2 + \tfrac{1}{2}kx^2 = \tfrac{1}{2}kA^2.$$

This equation can be used to find (a) the total energy of the body (6.25 J), (b) the amplitude of oscillation (0.35 m), (c) the velocity when the displacement is half of the amplitude (± 4.34 m/s,) (d) the displacement when the velocity is half of its initial value (± 0.31 m), and (e) the displacement when the kinetic and potential energies are equal (0.25 m).

The period and frequency are found from the relationships

$$f = \frac{1}{2\pi}\sqrt{\frac{k}{m}}, \qquad T = \frac{1}{f}.$$

The period is 0.44 s and the frequency is 2.25 Hz.

EVALUATE This problem illustrates how to apply simple harmonic energy relations to find displacements, velocities, and the amplitude, period, and frequency of the motion.

2: SHM practice

A body in SHM with angular frequency 0.5 s is initially 10.0 cm from its equilibrium position and is moving back toward equilibrium with a velocity of 5.0 cm/s, as shown in Figure 13.2. How long does it take for the body to return to its equilibrium position?

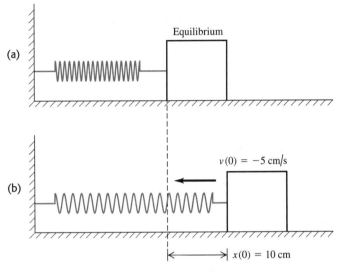

Figure 13.2 Try It Yourself 2.

Solution Checkpoints

IDENTIFY AND SET UP The body is in simple harmonic motion. Find the motion equation in terms of amplitude, angular frequency, and phase angle. Then solve for the time needed to move back to equilibrium, the target variable.

EXECUTE The general forms of the position and velocity equations are

$$x = A\cos(\omega t + \theta_0),$$
$$v = -\omega A\sin(\omega t + \theta_0).$$

To find the two constants, we use the initial conditions. At time zero, the position and velocity are given. This yields

$$A = 14.1 \text{ cm},$$
$$\theta_0 = 0.79 \text{ rad}.$$

The position is solved for time when $x = 0$, giving a time of 1.56 s.

EVALUATE This problem shows how to use initial conditions to solve for the equations of motion.

CAUTION **Watch f and ω!** Be careful to distinguish the frequency f from the angular frequency $\omega = 2\pi f$. The last two problems involved both quantities.

3: Simple pendulum

A clock pendulum with mass 5.0 kg is set to swing with a 2.0 s period. How long should the pendulum be made if you approximate it as a simple pendulum?

Solution Checkpoints

IDENTIFY AND SET UP The period of a simple pendulum is given in terms of the length of the pendulum and the gravitational constant.

EXECUTE The period of a simple pendulum is given by

$$T = 2\pi\sqrt{\frac{L}{g}}.$$

Rearranging terms and solving yields a length of 0.99 m.

EVALUATE Mass does not affect the results for a simple pendulum.

4: Physical pendulum

A body of mass 2.0 kg is suspended at a point 3.0 cm from its center of mass and observed to oscillate with a 2.0 s period. Find its moment of inertia.

Solution Checkpoints

IDENTIFY AND SET UP Is this a physical pendulum?

EXECUTE The period of a physical pendulum is given by

$$T = 2\pi\sqrt{\frac{I}{mgd}}.$$

Rearranging terms and solving yields a moment of inertia of 5.96×10^{-2} kg m^2.

EVALUATE This problem shows another method of determining the moment of inertia: Set the body in oscillation and measure the period.

14 Fluid Mechanics

Summary

We interact with fluids on a continual basis, from walking through air to swimming in the ocean. This chapter examines fluids, or substances that can flow, including liquids and gases. We will begin with fluid statics and use Newton's laws to describe the behavior of fluids at rest. Density, pressure, buoyancy, and surface tension, concepts needed for our investigation, will be defined. We will also delve into fluid dynamics and see how to analyze fluids in motion. Conservation of energy and Newton's laws will guide us in this examination. Although fluid dynamics can be quite complex, several examples will give us insight into the subject.

Objectives

After studying this chapter, you will understand

- The definition of a material's density.
- The definition of pressure in a fluid and its measurement.
- How to analyze fluids in equilibrium and find the pressure at varying depths.
- Buoyancy and how to calculate the buoyancy acting on a body.
- How to compare and contrast laminar and turbulent fluid flow.
- How to apply Bernoulli's equation to fluid dynamics problems.

Concepts and Equations

Term	Description
Density	Density is the mass per unit volume of a material. For a homogeneous material with mass m and volume V, the density is $$\rho = \frac{m}{V}.$$ The SI unit of density is the kilogram per cubic meter $(1\ \text{kg/m}^3)$. The cgs unit is used to express density in grams per cubic centimeter $(1\ \text{gm/cm}^3 = 1000\ \text{kg/m}^3)$.
Pressure	The pressure p in a fluid is the normal force per unit area: $$p = \frac{dF_\perp}{dA}.$$ The SI unit of pressure is the pascal (Pa); $1\ \text{Pa} = 1\ \text{N/m}^2$. Also common are the bar $(10^5\ \text{Pa})$ and millibar $(10^2\ \text{bar})$.
Pressure in a Fluid	The pressure difference between two points in a fluid with uniform density ρ is proportional to the difference in elevations between the two points: $$p_2 - p_1 = -\rho g(y_2 - y_1).$$ Pascal's law states that the pressure applied to a fluid is transmitted through the fluid and depends only on depth.
Buoyant Force	Archimedes' principle states that when an object is immersed in a fluid, the fluid exerts an upward buoyant force on the object equal in magnitude to the weight of the fluid displaced by the object.
Fluid Flow	An ideal fluid is incompressible and has no viscosity. Conservation of mass requires that the amount of fluid flowing through a cross section of a tube per unit time be the same for all cross sections: $$\frac{\Delta V}{\Delta t} = A_1 v_1 = A_2 v_2.$$
Bernoulli's Equation	Bernoulli's equation relates the pressure p, flow speed v, and elevation y of an ideal fluid at any two points: $$p_1 + \rho g y_1 + \frac{1}{2}\rho v_1^2 = p_2 + \rho g y_2 + \frac{1}{2}\rho v_2^2 = \text{constant}.$$

Conceptual Questions

1: Ice in a glass

Two glasses are filled with water to the same level. In one glass, ice cubes float on the top. If the two glasses are made of the same material and have the same shape, how do their total weights compare?

Solution

IDENTIFY, SET UP, AND EXECUTE The glasses must have the same weight, since they are made of the same material and have the same shape. We solve the problem by comparing the mass of the water alone to the mass of the water-plus-ice mix.

Archimedes' principle states that an object will displace its own weight in a fluid. The volume of water displaced by the ice has the same weight as the ice; therefore, the water in one glass weighs the same as the water plus ice in the second glass.

EVALUATE The volume of the glass with the ice is greater than the volume of the glass without ice, but weight depends on *both* density *and* volume. As with any new physical principle, we need to develop our skills carefully and not jump to conclusions.

2: Energy in a hydraulic lift

A hydraulic lift is used to lift a car. The piston supporting the car has a cross-sectional area 100 times larger than the cross-sectional area of the piston driving the lift. The drive piston will therefore require a force 100 times smaller than the weight of the car to lift the car. Does this mean that energy conservation is violated?

Solution

IDENTIFY, SET UP, AND EXECUTE Pascal's law states that the pressure is the same at both pistons. The drive piston requires a small force to create the pressure that will lift the car. A large displacement in the drive piston creates a small displacement in the lift piston, due to the differences in areas. The amount of work done in moving the drive piston a long distance is equal to the work done by the lift piston moving a small distance (ignoring friction). The amounts of work are equivalent; energy conservation is not violated.

EVALUATE If you have ever operated a hydraulic jack to lift your car or a house, you should recall that you had to pump the jack several times to move a small distance. The work produced by your small force applied over a long distance was equivalent to the work done in lifting the object.

3: Race car spoilers

Why do race cars have spoilers, or wings, on their bodies?

Solution

IDENTIFY, SET UP, AND EXECUTE Spoilers are essentially inverted airplane wings. We've seen that airplane wings produce lift by reducing the pressure above the plane's wing. The inverted wing produces a downward force to help hold the race car on the pavement and maintain contact between the wheels and the road. The spoiler also helps stabilize the car as it moves around the track.

EVALUATE Spoiler design for race cars is critical: There is a careful balance between enough downward force to keep the car on the track and too much force that causes lost fuel economy and premature tire wear. Some race cars have downward forces of up to three times the force of gravity (i.e., they could operate on an upside-down track and not fall off). Many cars have spoilers; most serve only an aesthetic purpose and have no effect on the car's performance.

Problems

1: How much seawater in a tank

Seawater is stored under pressure in a tank of horizontal cross-sectional area 4.5 m². The pressure above the seawater in the tank is 7.2×10^5 Pa, and the pressure at the bottom of the tank is 1.2×10^6 Pa. What is the mass of the seawater in the tank?

Solution

IDENTIFY We will use the relations among pressure, density, and height to solve the problem. The target variable is the mass of the seawater.

SET UP The tank is sketched in Figure 14.1. We can find the mass by first finding the volume of the seawater and then multiplying by the density. To find the volume, we multiply the height by the cross-sectional area of the tank. To find the height, we use the variation in the pressures due to the height of seawater. We assume that the seawater is incompressible.

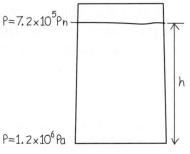

$P = 7.2 \times 10^5 \, Pn$

h

$P = 1.2 \times 10^6 \, Pa$

Figure 14.1 Problem 1 sketch.

EXECUTE We start by finding the height of the seawater in the container. The difference in pressure is related to the height by

$$p_2 - p_1 = \rho g (y_2 - y_1),$$

where we set p_1 as the pressure at the bottom of the tank and p_2 as the pressure at the top of the tank. Solving for the height, we get

$$h = y_2 - y_1 = \frac{p - p_0}{\rho g} = \frac{(1.2 \times 10^6 \, \text{Pa}) - (7.2 \times 10^5 \, \text{Pa})}{(1.03 \times 10^3 \, \text{kg/m}^3)(9.8 \, \text{m/s}^2)} = 47.6 \, \text{m},$$

where we used the density of seawater given in Table 14.1 in the text $(1.03 \times 10^3 \, \text{kg/m}^3)$. We now find the volume of seawater in the tank. The volume is the height times the cross-sectional area:

$$V = hA = (47.6 \, \text{m})(4.5 \, \text{m}^2) = 214 \, \text{m}^3.$$

The mass is the volume times the density we found from the table:

$$m = \rho V = (1.03 \times 10^3 \, \text{kg/m}^3)(214 \, \text{m}^3) = 221{,}000 \, \text{kg}.$$

The tank holds 221,000 kg of seawater.

EVALUATE This problem shows how to determine height from change in pressure. Altimeters find changes in altitude by monitoring the change in pressure of air.

2: Water pressure in a town

Water pressure in a town is maintained by a water tower 35.0 m high, open to the atmosphere at the top. (a) What is the gauge pressure at ground level? (b) A 1.75-cm-diameter garden hose at the bottom of the tower is open and spilling water. How much force is needed at the end of the hose to seal the end?

Solution

IDENTIFY We will use the relations among pressure, density, and height to solve the problem. The target variables are the gauge pressure at the ground and the force needed to seal the hose.

SET UP The gauge pressure is found by multiplying the height by the density by g, the acceleration due to gravity. The force is found by finding the pressure differential at the hose and multiplying by the area of the opening. We assume that the water is incompressible.

EXECUTE We start by finding the pressure at the surface. The gauge pressure is given by

$$p_g = p - p_a = \rho g (y_2 - y_1)$$
$$= (10^3 \text{ kg/m}^3)(9.8 \text{ m/s}^2)(35.0 \text{ m})$$
$$= 3.43 \times 10^5 \text{ Pa}$$
$$= 3.40 \text{ atm}.$$

The pressure difference between the inside and the outside of the hose is the gauge pressure, equivalent to the force per unit area needed to seal the hose. The force needed is then

$$F = (p - p_a)A = (3.43 \times 10^5 \text{ Pa})(\pi(0.0175/2 \text{ m})^2) = 83 \text{ N}.$$

The gauge pressure at the ground level is 3.40 atm, and the force required to seal the hose is 83 N.

EVALUATE This problem illustrates how force, pressure, and height are related in a fluid.

3: Velocity of water exiting a fire hose

Water enters a round fire hose of diameter 3.5 cm and exits from a round, 0.60-cm-diameter nozzle. If the water enters the hose at 2.0 m/s, what is the velocity of the exiting water? What is the maximum horizontal range of the water leaving the hose?

Solution

IDENTIFY The continuity equation relates the velocities and cross-sectional areas of incompressible fluids in a tube. The target variables are the velocity of the water at the outlet of the nozzle and the maximum range of the water.

SET UP We use the continuity equation to find the velocity of the water exiting the nozzle. To find the range, we employ projectile motion. We treat the water as incompressible.

EXECUTE The amount of fluid flowing through a tube per unit time is constant. The flow through the hose is equal to the flow through the nozzle:

$$A_{\text{hose}} v_{\text{hose}} = A_{\text{nozzle}} v_{\text{nozzle}}.$$

The area of the hose or nozzle is π times the square of half the diameter. Solving for the velocity of the nozzle gives

$$v_{\text{nozzle}} = \frac{A_{\text{hose}} v_{\text{hose}}}{A_{\text{nozzle}}} = \frac{\pi (D_{\text{hose}}/2)^2 v_{\text{hose}}}{\pi (D_{\text{nozzle}}/2)^2} = \frac{(3.5 \text{ cm})^2 (2.0 \text{ m/s})}{(0.6 \text{ cm})^2} = 68 \text{ m/s},$$

where the factors π and 2 cancel. The water molecules leaving the hose have a velocity of 68 m/s and undergo acceleration due to gravity. We use the kinematic relations for two-dimensional motion to find the range. Recall that the horizontal range of a projectile in terms of the launch angle θ_0 and the initial velocity v_0 is

$$R = \frac{v_0^2 \sin 2\theta_0}{g}.$$

The maximum range occurs when the launch angle is 45°. Substituting our values, we obtain

$$R = \frac{v_0^2 \sin 2(45°)}{g} = \frac{(68 \text{ m/s})^2 (1)}{(9.8 \text{ m/s}^2)} = 470 \text{ m}.$$

The maximum horizontal range of the water leaving the nozzle at 68 m/s is 470 m.

EVALUATE This problem illustrates why nozzles are placed at the ends of hoses. The reduced diameter of the nozzle increases the exit velocity and thereby increases the range of the water. Next time you wash your car, compare the velocity and range of the water leaving the hose with and without the nozzle attached.

4: Ice cube in glycerine

What fraction of an ice cube is submerged when floating in glycerine?

Solution

IDENTIFY We will use Newton's law and the definition of buoyancy to solve the problem. The target variable is the fraction of the ice cube submerged.

SET UP The buoyant force acts upward and gravity acts downward. The ice cube is in equilibrium, so the forces sum to zero.

EXECUTE The weight of the ice cube is

$$w_w = \rho_w g V_w.$$

The buoyant force is equal to the amount of glycerine displaced by the ice cube, which is equal to the weight of the ice cube. Combining terms yields

$$F_B = \rho_g g V_g = \rho_w g V_w.$$

The fraction of the ice cube that is submerged is the volume of glycerine displaced divided by the volume of the ice cube. Rearranging terms in the previous equation yields the fraction

$$\frac{V_g}{V_w} = \frac{\rho_w}{\rho_g} = \left(\frac{0.92 \text{ g/cm}^2}{1.26 \text{ g/cm}^2}\right) = 0.73.$$

Thus, 73% of the ice cube is submerged when floating in glycerine. Note that the densities of ice and glycerine were taken from Table 14.1 in the text.

EVALUATE We see how we can determine the fraction of ice located below the surface when the ice is placed in a liquid. You can use the same procedure to find out how much of an iceberg is below the surface in the ocean.

Practice Problem: Draw the free-body diagram of the problem.

5: Examining the buoyant force

A 4.5-cm-radius sphere of wood is held in fresh water below the surface by a spring. If the spring's force constant is 55 N/m, by how much is the spring stretched from its equilibrium position? Take the density of wood to be 700 kg/m^3.

Solution

IDENTIFY We will use Newton's law and the definition of buoyancy to solve for the stretch of the spring, the target variable. The wood is in equilibrium.

SET UP The free-body diagram of the block of wood is shown in Figure 14.2. The buoyant force is directed upward, and both gravity and tension due to the spring are directed downward. The block of wood is in equilibrium, so the forces sum to zero. The size and density of the wood determine its volume and mass, needed for the buoyant force and gravity. Hooke's law will be used to determine the amount of stretch.

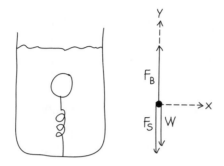

Figure 14.2 Problem 5 sketch and free-body diagram.

EXECUTE We sum the three forces acting on the block of wood. The forces act only in the vertical direction and add to zero:

$$\sum F_y = F_B - mg - F_s = 0.$$

The buoyant force is equal to the amount of water displaced by the wooden sphere. The volume of a sphere is $4/3\pi r^2$. Combining terms yields

$$F_B = \rho_{water} V_{sphere} g = \rho_{water} \left(\frac{4}{3} \pi r_{sphere}^3 \right) g.$$

The spring force is equal to the spring constant times its displacement, kx. The mass of the sphere is its volume times its density. Inserting these expressions into the equilibrium equation gives

$$\sum F_y = F_B - mg - F_s = \rho_{water} \left(\frac{4}{3} \pi r_{sphere}^3 \right) g - \rho_{wood} \left(\frac{4}{3} \pi r_{sphere}^3 \right) g - kx = 0.$$

Rearranging terms, we solve for x:

$$x = \frac{\rho_{water} \left(\frac{4}{3} \pi r_{sphere}^3 \right) g - \rho_{wood} \left(\frac{4}{3} \pi r_{sphere}^3 \right) g}{k} = \frac{(\rho_{water} - \rho_{wood}) \left(\frac{4}{3} \pi r_{sphere}^3 \right) g}{k},$$

$$x = \frac{((1 \times 10^3 \text{ kg/m}^3) - (700 \text{ kg/m}^3)) \left(\frac{4}{3} \pi (0.045 \text{ m})^3 \right) (9.8 \text{ m/s}^2)}{(55 \text{ N/m})} = 0.020 \text{ m}.$$

The spring is stretched 0.020 m, or 2.0 cm, from its equilibrium position.

EVALUATE This problem illustrates how to incorporate the buoyant force with previously encountered forces to solve equilibrium problems. We've used the same procedure as in the past: starting with a free-body diagram and setting the net force equal to zero. Working with fluids requires conversions among volume, mass, and density.

6: Water from a tank

A 10.0-m-high cylindrical tank of cross-sectional area 0.75 m² is filled with water. (a) Find the velocity of discharge as a function of the height of water remaining in the tank when a hole of area 0.40 m² is opened at the bottom of the tank. (b) Find the initial discharge velocity. (c) Find the initial volume rate of discharge.

Solution

IDENTIFY Bernoulli's equation and the continuity equation will be used to relate the pressure difference, height, and velocity of the flowing water. The target variables are the discharge velocity and volume rate of discharge.

SET UP Bernoulli's equation will be used to relate the velocities at the top and bottom of the tank to the change in height of the water. The continuity equation also relates the two velocities. Combining both equations will yield the velocity at the bottom of the tank, from which we can determine the solutions to parts (b) and (c).

EXECUTE Bernoulli's equation applied to the top and bottom of the cylinder gives

$$p_{top} + \rho g y_{top} + \frac{1}{2}\rho v_{top}^2 = p_{bottom} + \rho g y_{bottom} + \frac{1}{2}\rho v_{bottom}^2.$$

Both sides are open to atmospheric pressure, so the pressures are the same. We derive an expression for the velocities:

$$v_{bottom}^2 - v_{top}^2 = 2g(y_{top} - y_{bottom}) = 2gh.$$

The continuity equation yields another relation between the two velocities:

$$v_{bottom}A_{bottom} = v_{top}A_{top}$$

or

$$v_{top} = \frac{A_{bottom}}{A_{top}}v_{bottom}.$$

Placing the right-hand side of the latter equation into the previous equation and solving for the velocity at the bottom gives

$$v_{bottom}^2 - \left(\frac{A_{bottom}}{A_{top}}v_{bottom}\right)^2 = 2gh,$$

or

$$v_{bottom} = \sqrt{2gh}\left(1 - \left(\frac{A_{bottom}}{A_{top}}\right)^2\right)^{-1/2}.$$

The discharge velocity as a function of the height of the water remaining in the tank is

$$v_{bottom} = 1.20\sqrt{gh}.$$

The initial discharge velocity is when the tank begins to drain ($h = 10.0$ m) and is equal to 11.8 m/s. The discharge rate when the tank begins to drain is

$$\text{volume discharge rate} = A_{bottom}v_{bottom} = 4.73 \text{ m}^3/\text{s}.$$

EVALUATE We must carefully check the units in this problem. Do they cancel correctly ?

7: Lift on a car on a highway

As a car travels down the highway, the speed of the air flowing over the top of the car is higher than the speed of the air flowing under the car, thus creating lift. Estimate the lift on a car as it travels at 100 kph.

Take the density of air to be 1.20 kg/m^3, the car's area to be 6 m^2, and the height of the car to be 1.0 m. Assume that air travels under the car at 100 kph and over the top of the car at 140 kph.

Solution

IDENTIFY Bernoulli's equation gives the pressure difference between the top and bottom of the car. We will use that equation to find the target variable, the lift on the car.

SET UP We find the lift force acting on the car from the definition of pressure as force per unit surface area.

EXECUTE The car is moving through a fluid (air), so Bernoulli's equation can be applied. We'll compare the pressures below and above the car to find the pressure difference. Bernoulli's equation is

$$p_{above} + \rho g y_{above} + \frac{1}{2}\rho v_{above}^2 = p_{below} + \rho g y_{below} + \frac{1}{2}\rho v_{below}^2.$$

The pressure difference is then

$$\Delta p = \rho g y_{above} + \frac{1}{2}\rho v_{above}^2 - \frac{1}{2}\rho v_{below}^2,$$

where we have set the origin below the car $(y_{below} = 0)$. The pressure difference is

$$\Delta p = \rho \left(g y_{above} + \frac{1}{2}(v_{above}^2 - v_{below}^2) \right)$$

$$= (1.20 \text{ kg/m}^3)\left((9.8 \text{ m/s}^2)(1.0 \text{ m}) + \frac{1}{2}((38.9 \text{ m/s})^2 - (27.7 \text{ m/s})^2) \right)$$

$$= 459 \text{ Pa},$$

where we replaced 100 kpm with 27.7 m/s and 140 kph with 38.9 m/s. We find the force by multiplying the pressure by the area of the car:

$$F = PA = (459 \text{ Pa})(6.0 \text{ m}^2) = 2750 \text{ N}.$$

The lift on the car is 2750 N.

EVALUATE We see that the lift is significant in this case—roughly equivalent to a weight of 280 kg. It is not enough to lift the car off the highway, since most cars weigh over 1000 kg. We assumed that the flow of air around the car was smooth and that air is incompressible. Neither are valid assumptions and should be modified in a careful examination. Our results show the maximum lift of the car.

8: Pressure in a water system

Water is discharged from a closed system, reaching a maximum height of 10.0 m. What is the gauge pressure of the water system at the hose nozzle?

Solution

IDENTIFY Bernoulli's equation gives the pressure difference between the tank and the top of water in flight. The target variable is the gauge pressure in the tank.

SET UP Kinematics is used to find the initial velocity of the water, taking into account the maximum height. Bernoulli's equation is then used to find the pressure in the tank.

EXECUTE From kinematics, for the water to reach a height h, it must have an initial velocity of

$$v_{noz}^2 = 2gh.$$

Bernoulli's equation is applied to two points, one inside the tank and one at the nozzle, both at the same height, to find the pressure:

$$p_{in} + \rho gy + \frac{1}{2}\rho v_{in}^2 = p_{noz} + \rho gy + \frac{1}{2}\rho v_{noz}^2.$$

The outside pressure is atmospheric pressure, so the difference between the two pressures is the gauge pressure, our target variable. The velocity inside the tank is zero, giving

$$p_{gauge} = p_{in} - p_a = \tfrac{1}{2}\rho v_{noz}^2.$$

Combining the results produces

$$p_{gauge} = \frac{1}{2}\rho(2gh) = (10^3 \text{ kg/m}^3)(9.8 \text{ m/s}^2)(10.0 \text{ m}) = 9.8 \times 10^4 \text{ Pa}.$$

The tank is at a gauge pressure of 9.8×10^4 Pa.

EVALUATE We've combined kinematics and fluid dynamics in this problem. We'll continue building our physics models and call upon older material to help us as we move forward.

Try It Yourself!

1: Leak in a submarine

A submarine descends 35.0 m into the ocean and springs a leak. The hole out of which water is leaking has a diameter of 2.5 cm. What force must be used to plug the hole?

Solution Checkpoints

IDENTIFY AND SET UP We will use the relations among pressure, density, and depth to solve the problem. The target variable is the force needed to seal the hole. We must find the pressure at depth first. The density of seawater is $1.03 \times 10^3 \text{ kg/m}^3$.

EXECUTE The pressure difference between the water and the submarine (assuming that the latter is at a pressure of 1 atmosphere inside) is given by

$$p - p_a = \rho g(y_2 - y_1).$$

This equation yields a pressure difference of 3.50 atm. The force needed to plug the hole is

$$F = (p - p_a)A.$$

The force required to seal the hole is 173 N.

EVALUATE Do you think it is possible for someone to push with a force of 173 N and seal the hole?

2: Floating dumpling

A dumpling floats two-thirds submerged in water. What is its density?

Solution Checkpoints

IDENTIFY AND SET UP The dumpling is in equilibrium. What forces are acting on it? How do you quantify the buoyant force?

EXECUTE The fraction of the dumpling that is submerged is given by

$$\frac{V_D}{V_w} = \frac{\rho_w}{\rho_D}.$$

This equation can be rearranged to solve for the density of the dumpling. Doing so yields a density of 0.66 g/cm^2.

EVALUATE This problem illustrates how we can find the density of an object by examining how it floats.

3: Weighing a sphere under water

A sphere of volume 10 cm^3 displaces a spring scale and is found to weigh 50.0 g when submerged in water. What are the sphere's mass and density?

Solution Checkpoints

IDENTIFY AND SET UP The sphere is in equilibrium. What three forces are acting on it?

EXECUTE The net force acting on the sphere is

$$\sum F_y = F_s + F_B - mg = 0.$$

The weight and buoyant force are written in terms of density, volume, and g, the acceleration due to gravity. These quantities can be rearranged to yield

$$\rho = \frac{F_s}{Vg} + \rho_w.$$

From this equation, the density of the sphere is 6.0 g/cm^3 and the mass of the sphere is 60.0 g.

EVALUATE How did you decide the direction that the spring force acted in?

4: Water from a tank

Water inside an enclosed tank is subjected to a pressure of two atmospheres at the top of the tank. What is the velocity of discharge from a small hole 3.0 m below the surface of the water?

Solution Checkpoints

IDENTIFY AND SET UP Use Bernoulli's equation and the continuity equation to solve this problem. The target variable is the discharge velocity. The small hole indicates that the ratio of the hole to the surface area is small.

EXECUTE Bernoulli's equation applied to the top of the water and hole gives

$$p_{top} + \rho g y_{top} + \frac{1}{2}\rho v_{top}^2 = p_{hole} + \rho g y_{hole} + \frac{1}{2}\rho v_{hole}^2.$$

What is the pressure outside and at the top? Can you assume that the velocity at the top surface is small? The continuity equation indicates that

$$v_{top} = \frac{A_{hole}}{A_{top}} v_{hole} \approx 0.$$

Solving for the velocity at the hole gives

$$v_{hole}^2 = \frac{2}{\rho}(p_a + \rho g h),$$

or a velocity of 16 m/s.

EVALUATE How can you check these results?

5: Water from a rocket

A toy rocket of diameter 2.0 in consists of water under the pressure of compressed air pumped into the nose chamber. When the gauge air pressure is 60 lb/in^2, the water is ejected through a hole of diameter 0.2 in. Find the propelling force, or thrust, of the rocket.

Solution Checkpoints

IDENTIFY AND SET UP Use Bernoulli's equation and the continuity equation to solve this problem. The target variable is the thrust.

EXECUTE Bernoulli's equation applied to points inside and outside of the rocket, at the same height, gives

$$p_{in} + \rho g y + \frac{1}{2}\rho v_{in}^2 = p_{out} + \rho g y + \frac{1}{2}\rho v_{out}^2.$$

What is the pressure p_a outside the rocket? What is the velocity of water inside the rocket? The thrust is given by

$$F = v_{out}\frac{dm_{out}}{dt} = v_{out}\rho\frac{dV_{out}}{dt} = v_{out}\rho A_{out}v_{out}.$$

Solving for the force, we find that it is 3.8 lb.

EVALUATE Did you check units?

Mechanical Waves

Summary

In this chapter, we expand the concept of the periodic motion of an object to the periodic motion of many particles connected together as a medium. The periodic motion of a medium is a mechanical wave. Waves occur in many forms, including ocean waves, sound, light, earthquakes, and television transmission. This chapter will form the foundation for studying a variety of waves. We'll begin with the description of transverse and longitudinal waves and their amplitudes, periods, frequencies, and wavelengths. We'll see how waves move, interact, transmit energy, reflect, and combine in a variety of ways and how to describe their frequencies.

Objectives

After studying this chapter, you will understand

- How to identify longitudinal and transverse waves and their media.
- The relations among the period, velocity, frequency, and wavelength of a wave.
- How a wave function that satisfies the wave equation describes a wave.
- The concepts of superposition, standing waves, nodes, and antinodes.
- The allowed frequencies for standing waves.
- How waves interact and interfere.

Concepts and Equations

Term	Description
Mechanical Wave	A mechanical wave is a disturbance from equilibrium that propagates from one region of space to another through a medium. In a transverse wave, the particles in the medium are displaced perpendicular to the direction of travel. In a longitudinal wave, the particles in the medium are displaced parallel to the direction of travel.
Periodic Mechanical Waves	In a periodic wave, particles in the medium exhibit periodic motion. The speed, wavelength, period, and frequency of a periodic wave are related by $$v = \lambda f = \frac{\lambda}{T}.$$ The speed of a transverse wave in a string under tension is given by $$v = \sqrt{\frac{F_T}{\mu}},$$ where F_T is the tension in the rope and μ is the mass per unit length.
Wave functions	The wave function $y(x,t)$ describes the displacements of individual particles in the medium. For a sinusoidal wave traveling in the $+x$ direction, the equation $$y(x,t) = A\cos\left[\omega\left(\frac{x}{v} - t\right)\right]$$ $$= A\cos(kx - \omega t)$$ describes the wave. The wave functions are solutions of the wave equation $$\frac{\partial^2 y(x,t)}{\partial x^2} = \frac{1}{v^2}\frac{\partial^2 y(x,t)}{\partial t^2}.$$
Wave Power	Waves convey energy from one region to another. The average power of a sinusoidal wave is proportional to the squares of the wave frequency and amplitude: $$P_{av} = \frac{1}{2}\sqrt{\mu F}\,\omega^2 A^2.$$ As a wave spreads out into three dimensions, the intensity drop is inversely proportional to the distance from the source: $$\frac{I_1}{I_2} = \frac{r_2^2}{r_1^2}.$$
Principle of Superposition	The principle of superposition states that when two waves overlap, the net displacement at any point at any time is found by taking the sum of the displacements of the individual waves: $$y(x,t) = y_1(x,t) + y_2(x,t).$$
Standing Waves	A standing wave is that combination of sinusoidal waves which produces a stationary sinusoidal pattern. Nodes are points where the standing wave pattern does not change with time. Antinodes are positions halfway between nodes; the amplitude is maximum at an antinode. The distance between successive nodes or antinodes is one-half of the wavelength. A string of length L held stationary at both ends can have standing waves only with frequencies such that $$f_n = n\frac{v}{2L} = nf_1 \qquad (n = 1, 2, 3, \ldots).$$

The fundamental frequency is given by

$$f_1 = \frac{1}{2L}\sqrt{\frac{F}{\mu}}.$$

The multiples of f_1 are the harmonics. Each frequency and its associated vibration pattern is called a normal mode.

Conceptual Questions

1: Waves in a jump rope

Your younger sister is playing with a jump rope. She ties one end to a fence post and moves the other end up and down, observing waves in the rope. She sees waves moving down the rope and becomes confused. She asks you why the rope doesn't move toward the fence, since that is how the waves move. How do you answer her question?

Solution

IDENTIFY, SET UP, AND EXECUTE Having just learned about mechanical waves, you explain that the rope is made up of lots of little pieces (particles) and the pieces at the end she holds move up and down as she moves her hand up and down. The pieces of rope next to her end are connected to the pieces she is moving and so are pulled up and down at the same time. These pieces, which take a little bit longer to move than the ones she holds, are connected to other pieces, which also move up and down. Each successive piece takes a bit longer to start moving than the piece before it, which is why there is a wave pattern. This wave pattern is what your sister observes moving down the rope. All of the individual pieces of rope move only up and down. Since they are connected to each other, their moving creates a disturbance in the rope that appears to move down the rope. The rope doesn't move toward the fence because none of the pieces of the rope move toward the fence.

EVALUATE Remember that when you pluck a string, you pull the string to the side, so you give the string a velocity to the side of, or perpendicular to, the string. During the pluck, you impart a force *perpendicular* to the string, not along the string.

2: Velocity of a wave

A taut string is plucked and a wave travels down the string at speed v. How can you double the speed of the wave?

Solution

IDENTIFY, SET UP, AND EXECUTE The speed of a transverse wave on a string is proportional to the square root of the tension on the string and inversely proportional to the square root of the mass per unit length. To double the speed, you can quadruple the tension in the string or decrease the mass per unit length by a factor of 4.

EVALUATE This problem illustrates the dependence of the speed of a wave in a string on the mass per length and the tension in the string.

3: Heavy rope

A heavy rope is suspended vertically and is stretched taut by a 10.0-kg mass attached to the bottom of the rope. The top end of the rope is plucked, creating a wave. Does the speed of the wave change as it propagates down the rope? If so, how?

Solution

IDENTIFY, SET UP, AND EXECUTE The speed of a transverse wave on a string depends on the tension and the mass per unit length in the rope. We can assume that the mass per unit length is constant. Because the rope is heavy, the top end must have greater tension than the bottom end, since the top end supports both the mass at the end and the rope itself. The tension, therefore, decreases toward the bottom of the rope. The decreasing tension will slow the speed of the wave as it travels down the rope.

  The wave speed is not constant and slows as it approaches the bottom of the rope.

EVALUATE We will normally not encounter the varying tension and speed found in this problem. We'll focus on light strings to understand wave propagation better.

Problems

1: An unusual scale

Your strange physics professor builds an unusual scale by hanging an object from a 3.0-m-long wire attached to the ceiling. She plucks the string just above the object and finds that the pulse takes 0.50 s to propagate up and down the wire. What is the mass of the object? The mass of the wire is 0.50 kg.

Solution

IDENTIFY AND SET UP The speed of a wave in a wire under tension is related to the tension in the wire. By finding the speed of the wave in the wire, we'll determine the mass of the object.

EXECUTE The speed of the wave in the wire is given by

$$v = \sqrt{\frac{F_T}{\mu}}.$$

The tension at the bottom of the wire is equal to the gravitational force on the object, since the object is in equilibrium. To calculate the tension, we first need the velocity and the mass per unit length. The velocity is found by noting that the wave takes 0.50 s to travel 6.0 m (up and down the wire):

$$v = \frac{\Delta d}{\Delta t} = \frac{6.0 \text{ m}}{0.5 \text{ s}} = 12 \text{ m/s}.$$

The mass per unit length is

$$\mu = \frac{m}{L} = \frac{0.50 \text{ kg}}{3.0 \text{ m}} = 0.167 \text{ kg/m}.$$

Substituting to find the tension, we obtain

$$F_T = v^2\mu = (12 \text{ m/s})^2(0.167 \text{ kg/m}) = 24 \text{ N}.$$

The tension force is equal to the weight, so the mass of the object is

$$m = \frac{F_T}{g} = \frac{24 \text{ N}}{9.8 \text{ m/s}^2} = 2.4 \text{ kg}.$$

The object's mass is 2.4 kg.

EVALUATE This unusual scale illustrates how we can use mechanical waves to measure mass, but it is impractical for several reasons. First, the scale requires a high ceiling and a method of accurately measuring the speed of waves in the wire. Also, we have omitted the mass of the wire, which is roughly 15% higher at the ceiling, in our calculation of the tension.

2: Write a wave equation

Write the wave equation for a traveling transverse wave that propagates in the +x direction, has a maximum disturbance from equilibrium of 1.0 cm, and has a wavelength of 2.0 m and a period 0.02 s. At $x = 0.5$ m and $t = 0$, the instantaneous particle velocity is $\pi/2$ m/s downward.

Solution

IDENTIFY AND SET UP We'll begin with the general form of the wave function and use the given conditions to determine the constants.

EXECUTE The general form of a wave equation is

$$y(x,t) = A\sin(\omega t - kx + \phi).$$

Here, we used a minus sign in front of the wave number to ensure that the wave propagates toward positive x, and we included a phase angle. The frequency is found from the period:

$$\omega = 2\pi\frac{1}{T} = 314 \text{ rad/s}.$$

The wave number is related to the wavelength:

$$k = \frac{2\pi}{\lambda} = 3.14 \text{ /m}.$$

The amplitude is the maximum displacement from zero, so $A = 0.01$ m. To find the phase angle, we'll have to use the given velocity at the specified point. The velocity is the first derivative:

$$v_y = \frac{\partial y}{\partial t} = A\omega\cos(\omega t - kx + \phi).$$

At $x = 0.5$ m and $t = 0$, the instantaneous particle velocity is $\pi/2$ m/s downward, or negative. This gives

$$v_y(0.5 \text{ m}, 0) = -\pi/2 \text{ m/s}$$
$$(0.01 \text{ m})(314 \text{ rad/s})\cos((314 \text{ rad/s})(0) - (3.14 \text{ /m})(0.5 \text{ m}) + \phi) = -\pi/2 \text{ m/s}$$
$$(\pi)\cos(-\pi/2 + \phi) = -\pi/2$$
$$\sin(\phi) = -1/2.$$

The phase angle must be $-30°$, or $-\pi/6$. The complete wave function is

$$y(x,t) = (0.01 \text{ m})\sin((314 \text{ rad/s})t - (3.14 \text{ /m})x - \pi/6).$$

EVALUATE This problem illustrates how to construct a wave function, given the properties of the wave.

3: Wave function check

Does $y(x,t) = Ae^{-kx}\sin\omega t$ satisfy the wave function equation?

Solution

IDENTIFY AND SET UP We will take the second derivatives of the function with respect to both time and position. We will then substitute the results into the wave equation and see if it is satisfied.

EXECUTE The wave equation is given by

$$\frac{\partial^2 y(x,t)}{\partial x^2} = \frac{1}{v^2}\frac{\partial^2 y(x,t)}{\partial t^2}.$$

We start by taking the derivative of the given function with respect to position. We have

$$\frac{\partial y(x,t)}{\partial x} = \frac{\partial Ae^{-kx}\sin\omega t}{\partial x} = A(-k)e^{-kx}\sin\omega t.$$

Next, we take the second derivative with respect to position:

$$\frac{\partial^2 y(x,t)}{\partial x^2} = \frac{\partial}{\partial x}(A(-k)e^{-kx}\sin\omega t) = Ak^2 e^{-kx}\sin\omega t.$$

We now switch to the time derivatives. The first derivative with respect to time is

$$\frac{\partial y(x,t)}{\partial t} = \frac{\partial Ae^{-kx}\sin\omega t}{\partial t} = A(\omega)e^{-kx}\cos\omega t.$$

The second derivative is

$$\frac{\partial^2 y(x,t)}{\partial t^2} = \frac{\partial^2 Ae^{-kx}\sin\omega t}{\partial t^2} = A(-\omega^2)e^{-kx}\sin\omega t.$$

Combining the results gives

$$A(-\omega^2)e^{-kx}\sin\omega t = A(k^2)e^{-kx}\sin\omega t,$$

$$Ae^{-kx}\sin\omega t = -\frac{1}{v^2}Ae^{-kx}\sin\omega t.$$

The latter formula would satisfy the wave equation were it not for the minus sign. The function does not satisfy the wave equation and is not a valid wave function.

EVALUATE Constructing valid wave functions takes practice and experience. We'll see that there are several common types of equations that describe most waves and satisfy the wave equation.

4: Combining strings

Two strings of mass per unit length μ_1 and μ_2 are joined together at their ends. The tension in the two strings is the same. If the wavelength in the first string with $\mu_1 = 5.0$ g/m is 3.0 cm, what is the mass per unit length in the second string if the wavelength in that string is 5.0 cm? Assume that the wave frequencies are the same in the two strings.

Solution

IDENTIFY We'll use the fact that the tension is the same in both strings to determine the target variable, the mass per unit length in the second string.

SET UP The speed of propagation depends on the mass per unit length and the tension. The tension and frequency are the same in both strings. We'll set these equal to each other to solve.

EXECUTE The speed is given by

$$v = \sqrt{\frac{T}{\mu}}.$$

Squaring both sides of this equation and solving for the tension gives

$$v^2\mu = T.$$

Since both tensions are the same, we have

$$v_1^2\mu_1 = v_2^2\mu_2.$$

The speeds may not be the same, but we know the frequencies and wavelengths in both strings. The frequencies are the same in both strings. Rewriting the speed in terms of frequency and wavelength gives

$$\lambda_1^2 f^2 \mu_1 = \lambda_2^2 f^2 \mu_2.$$

Solving for the mass per unit length yields

$$\mu_2 = \frac{\lambda_1^2 \mu_1}{\lambda_2^2} = \frac{(3.0\text{ cm})^2(5.0\text{ g/m})}{(5.0\text{ cm})^2} = 1.8\text{ g/m}.$$

The second string has a mass per unit length of 1.8 g/m.

EVALUATE Do we expect a smaller mass per unit length in the second string? Yes, since the wavelength is larger in the second string, the mass per length must be less in order for the tensions to be the same.

Practice Problem: How do the speeds of the waves compare in the two strings? *Answer*: The speed of the wave in string 2 must be $\frac{5}{3}$ the speed in string 1.

5: Modes in a string

A uniform string of length 0.50 m is fixed at both ends. Find the wavelength of the fundamental mode of vibration. If the wave speed is 300 m/s, find the frequency of the fundamental and next possible modes.

Solution

IDENTIFY We'll use the properties and definitions of modes for a standing wave on a string to solve the problem.

SET UP The fundamental mode has a wavelength twice the length of the string. The frequency of the fundamental mode is the velocity divided by the wavelength. Higher frequencies are integer multiples of the fundamental frequency.

EXECUTE The wavelength of the fundamental mode is twice the length of the string. The wavelength is 1.0 m. For a wave speed of 300 m/s, the frequency of the fundamental mode is

$$f_1 = \frac{v}{2L} = 300 \text{ Hz.}$$

The next possible mode will have half of the fundamental mode's wavelength, or a wavelength of 0.5 m. Its frequency will be

$$f_2 = 2f_1 = \frac{v}{\lambda_2} = 600 \text{ Hz.}$$

EVALUATE This problem gives us practice understanding the properties of standing waves.

Try It Yourself!

1: Wave function check

Does $y(x,t) = A(x - vt)^n$ satisfy the wave function equation? A is a constant and $n > 1$.

Solution Checkpoints

IDENTIFY AND SET UP Take the second derivatives of the function with respect to both time and position. Then substitute the results into the wave equation and see if it is satisfied.

EXECUTE The second derivative with respect to position is

$$\frac{\partial^2 y(x,t)}{\partial x^2} = n(n - 1)A(x - vt)^{n-2}.$$

The second derivative with respect to time is

$$\frac{\partial^2 y(x,t)}{\partial t^2} = n(n - 1)Av^2(x - vt)^{n-2}.$$

Combining the results, we see that the function does satisfy the wave equation and is a valid wave function.

EVALUATE What type of wave does the function represent? Is it periodic?

2: Changing the diameter

Waves propagate through a rope under tension, provided by a hanging mass of 20.0 kg, with a speed of 30.0 m/s. The rope is replaced by ropes made of the same material but different diameters. For the velocity to remain the same, what mass should be hung from the end of the rope if the replacement rope has (a) half the diameter and (b) twice the diameter of the original rope?

Solution Checkpoints

IDENTIFY AND SET UP The speed of propagation depends on the tension and the mass per unit length.

EXECUTE How must the mass per unit length and the tension relate if the speed is to remain constant? How does the mass change with (a) half the diameter rope and (b) twice the diameter rope?

With half the diameter, the volume decreases by a factor of four, so the mass should be 5.0 kg. With twice the diameter, the volume increases by a factor of four, so the mass should be 80.0 kg.

EVALUATE How do you confirm these results?

3: Modes in a free string

A uniform string of length 0.50 m is fixed at one end and free at the other end. Find the wavelength of the fundamental mode of vibration. If the wave speed is 300 m/s, find the frequency of the fundamental and next possible modes.

Solution Checkpoints

IDENTIFY Use the properties and definitions of modes for a standing wave on a string to solve the problem.

SET UP How do the fundamental mode's frequency and wavelength for a standing wave on a string with only one end fixed compare with the fundamental mode's frequency and wavelength for a string with both ends fixed?

EXECUTE The wavelength of the fundamental mode is four times the length, or 2.0 m. The frequency of the fundamental mode is then 150 Hz.

The next mode has a wavelength of $\frac{4}{3}L$, or 0.67 m, and a frequency of 450 Hz.

EVALUATE Sketch the standing waves on the string for the fundamental and next possible modes. Does your sketch agree with the results you just calculated?

16 Sound and Hearing

Summary

In this chapter, we expand the concept of mechanical waves in order to understand sound and hearing. We'll begin with a description of longitudinal sound waves and their amplitudes, periods, frequencies, and wavelengths. We'll see how waves propagate through gases, liquids, and solids; how to determine the speed and intensity of sound waves; and how sound is produced by musical instruments. We'll also examine how the frequency of sound waves changes relative to the motion of the source and listener, summarized in the Doppler effect.

Objectives

After studying this chapter, you will understand

- How sound waves are formed and propagate through media.
- How to apply the concepts of superposition, standing waves, nodes, and antinodes to sound waves
- How to calculate the intensity of a sound wave.
- The allowed frequencies for longitudinal standing sound waves.
- The definition and how to calculate sound beats.
- How to apply the Doppler effect to moving sources and listeners.
- How to apply acoustics to a variety of systems.

Concepts and Equations

Term	Description
Sound Waves	Sound consists of longitudinal waves propagating through a medium. The pressure amplitude is given by $$p_{max} = BkA,$$ where B is the bulk modulus of the medium, k is the wave number, and A is the displacement amplitude. The speed of the sound wave depends on the medium: $$v = \sqrt{\frac{B}{\rho}} \quad \text{(longitudinal wave in a fluid)};$$ $$v = \sqrt{\frac{\gamma RT}{M}} \quad \text{(longitudinal wave in an ideal gas)};$$ $$v = \sqrt{\frac{Y}{\rho}} \quad \text{(longitudinal wave in a solid rod)}.$$
Intensity	The intensity of a sound wave is the rate at which energy is transported per unit area per unit time. The intensity of a sinusoidal wave is given by $$I = \frac{1}{2}\sqrt{\rho B}\,\omega^2 A^2 = \frac{p_{max}^2}{2\rho v} = \frac{p_{max}^2}{2\sqrt{\rho B}}.$$ The intensity β of a sound wave is a logarithmic measure given by $$\beta = (10\ \text{dB})\log\frac{I}{I_0},$$ where I_0 is the reference intensity (10^{-12} W/m^2). The units of β are decibels (dB).
Standing Sound Waves	Standing sound waves that propagate in a fluid in a pipe can reflect and form longitudinal standing waves. The closed end of a pipe is a displacement node and a pressure antinode; the open end of a pipe is a displacement antinode and a pressure node. For a pipe of length L with an open end, the fundamental frequency and harmonics are $$f_n = n\frac{v}{2L} = nf_1 \quad (n = 1, 2, 3, \dots).$$ For a pipe of length L with a closed end, the fundamental frequency and harmonics are $$f_n = n\frac{v}{4L} = nf_1 \quad (n = 1, 3, 5, \dots).$$
Interference	When waves overlap in the same region of space, the waves are said to interfere. When the waves combine to form a wave with a larger amplitude, the waves interfere constructively, or reinforce one another. When the waves differ by a half cycle, their sum results in a wave with a smaller amplitude, and the waves interfere destructively, or cancel.
Beats	Beats are heard when two tones of slightly different frequencies are sounded together, creating a beat frequency that is the difference of the original two frequencies. The beat frequency given by $$f_{beat} = f_a - f_b.$$

Doppler Effect	The Doppler effect is the frequency shift that occurs when the listener is in motion relative to the source of sound. The listener's frequency f_L is related to the source frequency f_S by $$f_L = \frac{v + v_L}{v + v_S} f_S,$$ where v is the speed of sound and v_L and v_S are the x components of the speed of the listener and source, respectively.

Conceptual Questions

1: Threshold of pain

By what factor must you amplify the intensity of a normal conversation to make it reach the threshold of pain?

Solution

IDENTIFY, SET UP, AND EXECUTE The intensity of a normal conversation is 65 dB, and the intensity at the threshold of pain is 120 dB, according to Table 16.2 in the text. These two intensities differ by 55 dB, or $10^{5.5} = 3.1 \times 10^5$. Therefore, you must amplify the normal conversation by 3.1×10^5 to reach the threshold of pain.

EVALUATE A normal conversation is five orders of magnitude less than the pain threshold. Remember this fact the next time your physics professor lectures: His lectures are *not* painful, because his voice hasn't reached the sound pain threshold!

2: Explaining Doppler shift

Your younger brother asks you to explain why the sound from train whistles changes from a high pitch to a low pitch when a train passes. How do you explain the change?

Solution

IDENTIFY, SET UP, AND EXECUTE You first explain that sound comes from vibrations, or changing pressure. A high pitch comes from faster vibrations and a low pitch comes from slower vibrations. The train whistle produces a steady number of vibrations. When the train approaches, the vibrations become compressed, effectively increasing the number of vibrations that reach your ear per unit time. When the train leaves, the vibrations become expanded, effectively slowing the vibrations.

EVALUATE This description provides an alternative explanation of the Doppler shift. As the listener and source move toward each other, the wave fronts become closer together and the frequency increases.

3: An orchestra warming up

While you wait for an orchestral performance, you hear the musicians tuning their instruments. You observe that several musicians play the same note for a few seconds to check the tune. What are they doing?

Solution

IDENTIFY, SET UP, AND EXECUTE The musicians are trying to play the same frequency when they tune their instruments. If one or more instruments vibrate at a slightly different frequency, the different frequencies interfere and produce beats. When tuning, the musicians listen for beats and readjust their instruments until the beats are removed.

EVALUATE You can listen for beats when you hear music. Beats may be intentional; for example, some pipe organs have a slow beat to create an undulating effect.

4: Frequencies in a water bottle

Estimate the two lowest frequencies that you can achieve by blowing across the top of a plastic water bottle.

Solution

IDENTIFY, SET UP, AND EXECUTE We can treat the bottle as a pipe that is open at one end and closed at the other. The normal-mode frequencies of a closed-end pipe are given by

$$f_n = \frac{nv}{4L},$$

where we want frequencies corresponding to $n = 1$ and 3. The water bottle is approximately 8", or 20.3 cm, long. Taking the speed of sound to be 344 m/s, we find that the two frequencies are 424 Hz and 1271 Hz.

EVALUATE Note that only the odd n's are valid for the closed-end pipe.

You can now build a pipe organ from recycled soda cans and soda bottles.

Problems

1: Finding a plug in a tube

In an attempt to find where a plug is in a tube containing air, a plumber blows air across the opening of the tube and hears a resonance at a frequency of 80 Hz. If this is the fundamental mode, how far away from the end of the pipe is the plug? Take the velocity of sound in air to be 345 m/s.

Solution

IDENTIFY We'll use the relationship between pipe length and normal-mode frequencies to find the distance the plug is from the end of the tube—the target variable.

SET UP We will use the normal-mode relationship to find the fundamental frequency of a closed pipe, since the plug effectively closes the pipe.

EXECUTE The fundamental frequency of the closed pipe is

$$f_1 = \frac{v}{4L}.$$

We know the frequency and speed of sound, so we solve for the length:

$$L = \frac{v}{4f_1} = \frac{(345 \text{ m/s})}{4(80 \text{ Hz})} = 1.08 \text{ m}.$$

The plug is 1.08 from the end of the pipe.

EVALUATE This problem illustrates how to count overtones carefully and how to interpret integer results.

2: Overtones in a pipe

An open pipe of length 1.5 m is played on a day when the speed of sound in air is 345 m/s. How many overtones can be heard by a person with good hearing?

Solution

IDENTIFY The number of overtones is the number of frequencies above the fundamental frequency. We'll use the relationship between pipe length and normal-mode frequencies to find the number of overtones—the target variable.

SET UP A person with good hearing can hear in the range from 20 Hz to 20,000 Hz. We'll find the number of frequencies in that range for the pipe, and then we'll subtract the fundamental frequency to solve the problem.

EXECUTE The frequencies of standing waves in an open pipe are

$$f_n = nf_1,$$

where n is an integer. The fundamental frequency of the pipe is

$$f_1 = \frac{v}{2L}.$$

For this pipe,

$$f_1 = \frac{v}{2L} = \frac{(345 \text{ m/s})}{2(1.5 \text{ m})} = 115 \text{ Hz}.$$

The fundamental frequency is above 20 Hz, so it can be heard. The highest frequency that can be heard by the human ear is 20,000 Hz. This frequency corresponds to

$$n = \frac{20,000 \text{ Hz}}{f_1} = \frac{20,000 \text{ Hz}}{115 \text{ Hz}} = 173.9.$$

Since we cannot hear nine-tenths of a frequency, we truncate n to 173. Thus, 173 frequencies can be heard: the fundamental frequency and 172 overtones.

EVALUATE This problem illustrates how to count overtones carefully and how to interpret integer results.

3: Making notes

A 1-meter-long tube open at one end and closed at the other contains water to a depth d. Assuming that the sound waves have a displacement node at the water surface and an antinode at the open end, find the depth of liquid that makes the tube resonate at middle C (264 Hz) and one octave below middle C (132 Hz). Take the speed of sound in air to be 345 m/s.

Solution

IDENTIFY We'll use the relationship between pipe length and normal-mode frequencies to find the distance the water is from the top of the tube. We'll subtract that distance from the length of the pipe to find the height of the water—the target variable.

SET UP We will use the normal-mode relationship to find the fundamental frequency of a closed pipe, since there is a displacement node at the water. We will find the lowest mode, that corresponding to $n = 1$.

EXECUTE The fundamental frequency of a closed pipe is

$$f_1 = \frac{v}{4L},$$

where L is the distance from the top of the tube to the top of the water. We solve for L for the two frequencies:

$$L_{\text{middle C}} = \frac{v}{4f_1} = \frac{(345 \text{ m/s})}{4(264 \text{ Hz})} = 0.327 \text{ m};$$

$$L_{\text{below C}} = \frac{v}{4f_1} = \frac{(345 \text{ m/s})}{4(132 \text{ Hz})} = 0.653 \text{ m}.$$

The water must be at a height of $1.0 \text{ m} - 0.327 \text{ m} = 0.673 \text{ m}$ for middle C and at a height of $1.0 \text{ m} - 0.653 \text{ m} = 0.347 \text{ m}$ for one octave below middle C.

EVALUATE This problem illustrates how to design a pipe organ that is made by filling several pipes of the same length with water. The pipe, however, would be a tough organ to tune, because the water will evaporate.

4: Power at a concert

You are given the task of determining how much power is needed in the sound system at the new stadium. There is a single set of speakers on the stage. If the design calls for an intensity of 100 dB at the farthest seats (120 m from the speakers), how much power is required?

Solution

IDENTIFY AND SET UP In order to determine the intensity from the power, we assume that the sound is distributed over a sphere of radius 120 m. We'll use the definition of intensity to relate the design intensity to the power.

EXECUTE The intensity is given by

$$\beta = (10 \text{ dB})\log\frac{I}{I_0}.$$

Taking the logarithm of both sides gives

$$I = I_0 10^{(\beta/10 \text{ dB})}.$$

The intensity is the power per unit area, where the area in this case is that of a sphere $(4\pi r^2)$. Combining the various equations gives

$$P = AI = (4\pi r^2)(I_0 10^{(\beta/10 \text{ dB})}) = (4\pi(120 \text{ m})^2)((10^{-12} \text{ W/m}^2)10^{((120 \text{ dB})/10 \text{ dB})}) = 180 \text{ kW}.$$

The required power is 180 kW.

EVALUATE Our stadium requires a substantial sound system in order for all visitors to hear the concert. The amount of power required would harm the hearing of those near the speaker. Stadiums are designed with multiple speakers placed around the stadium and closer to the visitors, to reduce the maximum volume.

Practice Problem: What is the intensity of the sound for persons seated 20 m from the speakers? *Answer:* 135 dB, above the threshold for permanent hearing damage.

5: Speed of approaching train

You are driving along a country road at 20.0 m/s. A train approaches on a rail that parallels the road. The train whistle blasts at a frequency of 800 Hz, but you hear a 950-Hz whistle. What is the speed of the approaching train?

Solution

IDENTIFY Our target variable is the speed of the approaching train.

SET UP The Doppler effect describes the frequency shift for moving sources, so we'll use the Doppler formula to determine the speed of the approaching train. Our coordinate system is shown in Figure 16.1; positive velocities are taken to be from the listener toward the source. Both the source and listener are moving. The listener's velocity is positive and the source's velocity is negative in our coordinate system. The speed of sound is taken to be 340 m/s.

Figure 16.1 Problem 5 sketch.

EXECUTE The listener's frequency is related to the source frequency by the Doppler shift,

$$f_L = \frac{v + v_L}{v + v_S} f_S,$$

where v is the speed of sound and v_L and v_S are the x components of the speed of the listener and source, respectively. We can rearrange to solve for v_S:

$$v_S = \frac{v + v_L}{f_L} f_S - v.$$

In this case, v_L is $+20.0$ m/s, f_L is 950 Hz, f_S is 800 Hz, and v is 340 m/s. Substituting yields

$$v_S = \frac{v + v_L}{f_L} f_S - v = \frac{(340 \text{ m/s}) + (20.0 \text{ m/s})}{(950 \text{ Hz})} (800 \text{ Hz}) - (340 \text{ m/s}) = -36.8 \text{ m/s}.$$

The train is approaching at 36.8 m/s, or 132 kilometers per hour.

EVALUATE This problem illustrates how to use the Doppler shift to find the speed of an object. We expected and found a negative speed, indicating that the source was moving toward the listener. We see that proper Doppler-shift solutions require a coordinate system and careful interpretation of the directions of the velocities.

Practice Problem: What frequency would you hear if you were moving away from the train at 20.0 m/s? *Answer*: 844 Hz.

Try It Yourself!

1: Designing an organ pipe

You are asked to design an organ pipe that will produce a middle C on the "even-tempered scale." Middle C is equivalent to a frequency of 261.6 Hz. (a) If the tube is open at both ends, how long should it be? (b) If the tube is open at one end and closed at the other, how long should it be? Take the speed of sound in air to be 345 m/s.

Solution Checkpoints

IDENTIFY AND SET UP Use the normal-mode frequency relationships for open and closed tubes to find the lengths.

EXECUTE The fundamental frequency of an open pipe is

$$f_1 = \frac{v}{2L}.$$

The fundamental frequency of a closed pipe is

$$f_1 = \frac{v}{4L}.$$

The two lengths are 0.659 m and 0.330 m.

EVALUATE Which pipe is more desirable for a compact pipe organ?

2: Designing a organ pipe, part 2

Find the allowed normal-mode frequencies of the two organ pipes in the previous problem. Take the speed of sound in air to be 345 m/s.

Solution Checkpoints

IDENTIFY AND SET UP Use the normal-mode frequency relationships for open and closed tubes to find the frequencies.

EXECUTE The normal-mode frequencies of an open pipe are

$$f_n = \frac{nv}{2L}.$$

Substituting $n = 1, 2, 3 \ldots$, we find that $f_1 = 261.6$ Hz, $f_2 = 523.2$ Hz, $f_3 = 784.8$ Hz, ...
The fundamental frequency of a closed pipe is

$$f_1 = \frac{v}{4L}.$$

The normal-mode frequencies are integer multiples of the fundamental frequency; that is, $f_n = nf_1$. So, substituting $n = 1, 3, 5 \ldots$ we find that $f_1 = 261.6$ Hz, $f_2 = 784.8$ Hz, $f_3 = 1308$ Hz, ...

EVALUATE Which pipe is more desirable for the number of frequencies it can produce?

3: **Doppler-shift practice**

Consider a source that produces a sound with a frequency of 500 Hz. If the speed of sound in air is 345 m/s, and the source and listener both move along the line joining them at speeds of 25 m/s, what frequencies can be heard by the listener for all possible directions of velocities?

Solution Checkpoints

IDENTIFY AND SET UP Use the Doppler-shift equation to find the possible frequencies while varying the direction, or sign, of the velocities.

EXECUTE The listener's frequency is related to the source frequency by the Doppler shift,

$$f_L = \frac{v + v_L}{v + v_S} f_S,$$

where v is the speed of sound and v_L and v_S are the speeds of the listener and source, respectively. When the two velocities are in the same direction, either positive or negative, there is no Doppler shift and the listener hears a sound with a frequency of 500 Hz.

If the source and listener are moving away from each other, (say, v_L is negative and v_S is positive,) then the frequency is reduced to 432 Hz. If the source and listener are moving toward each other (say, v_L is positive and v_S is negative), then the frequency is increased to 579 Hz.

EVALUATE This problem illustrates the amount of Doppler shift for four possible combinations of velocity between two objects.

17 Temperature and Heat

Summary

In this chapter, we will begin a four-chapter investigation of thermo-dynamics. We lay the groundwork for the upcoming chapters with an initial definition of temperature, and then we see how materials change size with temperature. Heat will be introduced as a method of energy transfer due to temperature differences, and the rate of heat transfer will be calculated. We will also learn about the amount of heat required to change the phase of matter and about the three types of heat transfer: conduction, convection, and radiation.

Objectives

After studying this chapter, you will understand

- The definition of temperature and thermal equilibrium.
- The three temperature scales and how to measure temperature.
- How thermal expansion describes the change in length and volume of materials due to temperature changes.
- About heat, phase changes, and calorimetry and how to apply these concepts to problems.
- How heat is transferred by conduction, convection, and radiation.

Concepts and Equations

Term	Description
Thermal Equilibrium	Two objects in thermal equilibrium have the same temperature.
Temperature Scales	The Celsius temperature scale defines 0°C as the freezing point of water and 100°C as the boiling point of water. The Fahrenheit temperature scale defines 32°F as the freezing point of water and 212°F as the boiling point of water. The Kelvin scale defines absolute zero as 0 K and uses the Celsius unit as its standard unit. Note that 0 K is −273.15°C.
Thermal Expansion and Thermal Stress	Materials change size, or thermally expand, due to changes in temperature. An object of length L_0 at temperature T_0 will have length L at temperature $T = T_0 + \Delta T$, or $$L = L_0 + \Delta L = L_0(1 + \alpha \Delta T),$$ where a is the coefficient of linear expansion with units K^{-1}. An object of volume V_0 at temperature T_0 will have volume V at temperature $T = T_0 + \Delta T$, or $$V = V_0 + \Delta V = V_0(1 + \beta \Delta T),$$ where β is the coefficient of volume expansion with units K^{-1}. When a material is heated or cooled while being held such that it cannot contract or expand, it is under tensile stress given by $$\frac{F}{A} = -Y\alpha \Delta T.$$
Heat	Heat is energy transferred from one object to another due to changes in temperature. The quantity of heat Q needed to raise the temperature of a mass m of material by an amount ΔT is $$Q = mc\Delta T,$$ where c is the specific heat capacity of the material. The SI unit of heat capacity is the joule per kilogram per kelvin (J/(kg K)).
Phase Change	A phase transition is the change from one phase of matter to another. Phases include solid, liquid, and gas. The heat of fusion, L_f, is the heat per unit mass required to change a solid material to liquid. The heat of vaporization, L_v, is the heat per unit mass required to change a liquid material to gas. The heat of sublimation, L_s, is the heat per unit mass required to change a solid material to gas.
Calorimetry	Calorimetry is the measurement of heat in a system. For an isolated system, the algebraic sum of the quantities of heat must add to zero: $$\sum Q = 0.$$
Heat Transfer	Heat may be transferred through conduction, convection, and radiation. Conduction is the transfer of energy within a material without bulk motion of the material. Convection is the transfer of energy due to the motion of mass from one region to another. Radiation is the transfer of energy through electromagnetic waves. The heat current H for an area A and length L through which the heat flows is given by $$H = \frac{dQ}{dt} = kA\frac{T_H - T_C}{L},$$

where T_H and T_C are, respectively, the temperatures of the hot and cold sides of the material and k is the thermal conductivity. The heat current H due to radiation is

$$H = Ae\sigma T^4,$$

where A is the surface area, e is the emissivity of the surface (a pure number between 0 and 1), T is the absolute temperature, and σ is the Stefan-Boltzmann constant (5.6705×10^{-8} W/m^2/K^4).

Conceptual Questions

1: Do holes expand or contract?

Your younger brother knows that solids expand as they heat up. He thinks that the metal surrounding the hole in a cookie sheet will expand into the hole as the cookie sheet heats up. Is he right or wrong?

Solution

IDENTIFY, SET UP, AND EXECUTE Figure 17.1 shows a sketch of a cookie sheet with a hole in it. After considering the problem, you realize that the hole will enlarge as the cookie sheet heats up, since all dimensions of an object enlarge with temperature. The challenge is how to best explain this phenomenon to him.

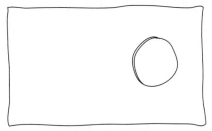

Figure 17.1 Question 1 sketch.

If you give the problem a bit more thought, you come up with a convincing argument. If the cookie sheet had no hole, the whole sheet would increase with temperature. If you punch out a hole in the same cookie sheet and consider the piece of metal that was removed, this piece expands as its temperature rises. Therefore, the hole in the cookie sheet must also expand, just as it did when the cookie sheet was holeless.

EVALUATE Thermal expansion must be considered carefully. Here, we see that a confusing point can be clarified by imagining what happens to the piece that was once the hole.

CAUTION **Holes expand when heated!** Keep the results of this problem in mind when you encounter similar problems. Holes don't shrink when heated.

2: Cooler after a shower

When you step out of the shower, you often feel cold. After drying off, you feel warmer, even though the room's temperature is the same as when you stepped out of the shower. Why?

Solution

IDENTIFY, SET UP, AND EXECUTE When you step out of the shower, water on your body evaporates. Evaporation requires heat energy (the heat of vaporization), much of which energy comes from heat leaving your body. You feel cold because your body is transferring its heat to evaporate the water. When you are dry, there is little heat lost due to evaporation.

EVALUATE Evaporation also explains why one feels cooler in a dry climate than in a humid climate: Your sweat evaporates more rapidly in a dry climate, taking away more heat, than in a humid climate.

3: Cold water versus cold air

Would you prefer to spend 10 minutes in a 40°F (4°C) room or in a 40°F pool? Why?

Solution

IDENTIFY, SET UP, AND EXECUTE Both the room and the pool are at the same temperature, but the 40°F room would be much more comfortable. The reason is that the specific heat of air is much less than the specific heat of water (i.e., air will carry away less heat from your body than the water would in any time interval). Since the air carries away less heat, you are more comfortable in the room.

EVALUATE Specific heat is the amount of heat needed to change the temperature of a material per unit mass and per unit temperature. Larger specific heats mean that more heat is carried away from an object.

Problems

1: Volume of a copper cup

A copper cup is filled to the brim with ethanol at 0°C. When the cup and ethanol are heated to 35°C, 4.7 cm³ of ethanol spills from the cup. What is the initial volume of the cup?

Solution

IDENTIFY We will use temperature expansion to find the change in volume of the cup and the ethanol. Their difference will lead to the initial volume of the cup—the target variable.

SET UP Both the cup and the ethanol expand as the temperature rises; the difference in their expansion is equal to the volume of the spilled ethanol. We'll apply the volume expansion equation to both the cup and the ethanol, setting their difference equal to the volume of the spill. The coefficient of volume expansion is $5.1 \times 10^{-5}/\text{K}$ for copper and $75 \times 10^{-5}/\text{K}$ for ethanol.

EXECUTE For any material, the change in volume due to temperature is

$$\Delta V = \beta V_0 \Delta T.$$

We are given the volume of the spill, which is the change in volume of the ethanol minus the change in volume of the cup:

$$V_{\text{spill}} = \Delta V_{\text{ethanol}} - \Delta V_{\text{cup}}.$$

The initial volumes of the cup and the ethanol are the same. We'll call their common volume V_0. The temperature of both materials is 35°C. Replacing the changes in volumes yields

$$V_{\text{spill}} = \beta_{\text{ethanol}}V_0\Delta T - \beta_{\text{copper}}V_0\Delta T.$$

Solving for V_0, we obtain

$$V_0 = \frac{V_{\text{spill}}}{(\beta_{\text{ethanol}} - \beta_{\text{copper}})\Delta T} = \frac{(4.7 \text{ cm}^3)}{((75 \times 10^{-5}/\text{K}) - (5.1 \times 10^{-5}/\text{K}))(35°\text{C})} = 190 \text{ cm}^3.$$

The original volume of the cup is 190 cm³.

EVALUATE This is a straightforward application of volume thermal expansion. We did need to note carefully that both the copper and the ethanol expanded, so the spillage was the difference in the changes in volumes. We could almost ignore the change in volume of the copper, since the coefficient of thermal expansion is much smaller for copper than for ethanol.

Practice Problem: What is the final volume of the cup? *Answer*: 190.3 cm³.

2: Stress in a wire

An aluminum wire is stretched across a large steel frame. Initially, the wire is at 20°C and is unstressed. The system (wire plus frame) is cooled by 50°C. If the area of contact for the wire is $9.0 \times 10^{-6} \text{ m}^2$, what force is exerted on the wire?

Solution

IDENTIFY The differences in the expansion of the wire and frame will lead to tensile stress acting on the wire. The target variable is the force on the wire.

SET UP We will first use temperature expansion to find the change in lengths of the wire and frame. We will then find the tensile stress on the wire. The coefficient of linear expansion is $2.4 \times 10^{-5}/\text{K}$ for aluminum and $1.2 \times 10^{-5}/\text{K}$ for steel. Young's modulus for the aluminum is $0.7 \times 10^{11} \text{ N/m}^2$.

EXECUTE The percent change in length for a material due to temperature is

$$\Delta L = \alpha L_0 \Delta T.$$

When the aluminum is cooled by 50°C, the change in its length is

$$\left(\frac{\Delta L}{L_0}\right)_{\text{al}} = \alpha_{\text{al}}\Delta T = (2.4 \times 10^{-5})(50) = 1.2 \times 10^{-3}.$$

For the steel, the change in length is

$$\left(\frac{\Delta L}{L_0}\right)_{\text{st}} = \alpha_{\text{st}}\Delta T = (1.2 \times 10^{-5})(50) = 0.60 \times 10^{-3}.$$

Both of these changes are decreases, since the temperature has decreased. We see the aluminum changes more than the steel. Because the steel decreases less, stress is induced in the aluminum. The stress is given by

$$\frac{F}{A} = Y_{\text{al}}\left(\frac{\Delta L}{L_0}\right).$$

The stress is proportional to the net change in length of the wire. The frame shrinks, so the net change in length is the difference in the changes of the wire and the frame. The force is then

$$F = AY_{al}\left(\frac{\Delta L}{L_0}\right) = (9.0 \times 10^{-6}\ \text{m}^2)(0.7 \times 10^{11}\ \text{N/m}^2)(1.2 \times 10^{-3} - 0.60 \times 10^{-3}) = 380\ \text{N}.$$

The stress on the wire is 380 N.

EVALUATE This is an application of linear thermal expansion. To find the stress, we did need to note carefully the difference in how the two materials contracted. We could also have used the relation in the book that directly provides the stress as a function of temperature.

3: Ice to steam

A copper calorimeter of mass 2.0 kg initially contains 1.5 kg of ice at $-10°C$. How much heat energy must be added to convert all of the ice to water and then half of the water into steam?

Solution

IDENTIFY We will use heat capacity and heat of fusion to determine how the ice melts and turns to steam. The target variable is the amount of heat needed to convert the ice to water and the water to steam.

SET UP We will solve the problem in several steps. We'll first find the heat required to raise the temperature of the ice to 0°C, then find the heat required to melt the ice, then find the heat required to warm the water to 100°C, and finally find the heat required to vaporize half of the water. We know that the final temperature of the remaining water must be 100°C, since the water remains in equilibrium throughout the process. We must also add the heat required to heat the copper pot to 100°C.

EXECUTE The heat required to heat the ice to 0°C is

$$Q_1 = m_{ice}c_{ice}\Delta T = (1.5\ \text{kg})(2100\ \text{J/kg/K})(10.0°C) = 31,500\ \text{J}.$$

We used the specific heat of ice (2010 J/kg/K) to find Q_1. The heat required to melt the ice is the heat of fusion for ice:

$$Q_2 = m_{ice}L_f = (1.5\ \text{kg})(3.34 \times 10^5\ \text{J/kg}) = 501,000\ \text{J}.$$

The melted ice must warm to the final temperature (100.0°C):

$$Q_3 = m_{ice}c_{water}\Delta T = m_{ice}(4190\ \text{J/kg/K})(100.0°C) = 628,500\ \text{J}.$$

Here, we used the heat capacity of water (4190 J/kg/K). Half of the mass turns to steam. Using the heat of vaporization $(2.256 \times 10^6\ \text{J/kg})$, we find that the heat required is

$$Q_4 = \tfrac{1}{2}m_{ice}L_f = \tfrac{1}{2}(1.5\ \text{kg})(2.256 \times 10^6\ \text{J/kg}) = 1,692,000\ \text{J}.$$

The copper pot also increases in temperature, from $-10°C$ to 100°C. The heat required to bring about this increase is

$$Q_{copper} = m_{copper}c_{copper}\Delta T = (2.0\ \text{kg})(390\ \text{J/kg/K})(110.0°C) = 85,800\ \text{J}.$$

The total heat is the sum of the five quantities of heat:

$$Q_t = Q_1 + Q_2 + Q_3 + Q_4 + Q_{copper} = 2.94 \times 10^6\ \text{J}.$$

So 2.94×10^6 J are needed to heat the pot and ice from $-10°C$ to 100°C and then to vaporize half of the water that is produced.

EVALUATE We see how we must solve calorimetry problems in multiple steps. We need to include the latent heat when materials change phase and the heat capacity when materials heat up.

CAUTION **Temperature is not heat!** Temperature characterizes the state of an object. Heat is the flow of energy. The two terms may be synonymous in everyday language, but they are not synonymous in physics.

4: Cooling hot tea

You wish to chill your freshly brewed tea with the minimum amount of ice that will avoid watering it down too much. What is the minimum amount of ice you should add to 2.0 kg of freshly brewed tea at 95°C to cool it to 5.0°C? The ice is initially at a temperature of −5.0°C.

Solution

IDENTIFY We'll set the heat lost from the tea equal to the heat gained by the ice. The target variable is the amount of ice needed to cool the tea.

SET UP The amount of heat lost by the tea is given by the specific heat capacity equation, since the tea doesn't go through a phase change. The ice melts, so, in calculating the heat gain of the ice, we need to include the latent heat of fusion, plus the changes due to the ice warming to 0°C, and the changes due to the melted ice warming to 5.0°C.

EXECUTE The heat transfer from the hot tea as it cools to 5.0°C is negative:

$$Q_{tea} = m_{tea}c_{water}\Delta T_{tea} = (2.0 \text{ kg})(4190 \text{ J/kg/K})(5.0°C - 95°C) = -754{,}000 \text{ J}.$$

Here, we used the heat capacity of water (4190 J/kg/K) for the tea. The ice must warm to 0°C, then melt, and then heat to 5.0°C. We find the heat required for each segment of the ice warming. For the ice to heat to 0°C, we use the specific heat of ice (2010 J/kg/K):

$$Q_{ice} = m_{ice}c_{ice}\Delta T_{ice} = m_{ice}(2010 \text{ J/kg/K})[0.0°C - (-5.0°C)] = m_{ice}(10{,}000 \text{ J/kg}).$$

The heat needed to melt the ice is the heat of fusion for ice:

$$Q_{melt} = m_{ice}L_f = m_{ice}(3.34 \times 10^5 \text{ J/kg}).$$

The melted ice must warm to the final temperature (5.0°C):

$$Q_{melted \ ice} = m_{ice}c_{water}\Delta T_{melted \ ice} = m_{ice}(4190 \text{ J/kg/K})(5.0°C - 0.0°C) = m_{ice}(21{,}000 \text{ J/kg}).$$

The sum of these four quantities must be zero:

$$Q_{tea} + Q_{ice} + Q_{melt} + Q_{melted \ ice} = -754{,}000 \text{ J} + m_{ice}(10{,}000 \text{ J/kg})$$
$$+ \ m_{ice}(334{,}000 \text{ J/kg}) + m_{ice}(21{,}000 \text{ J/kg}) = 0.$$

Therefore,

$$m_{ice} = \frac{754{,}000 \text{ J}}{(10{,}000 \text{ J/kg}) + (334{,}000 \text{ J/kg}) + (21{,}000 \text{ J/kg})} = 2.1 \text{ kg}.$$

It takes a minimum of 2.1 kg of ice to cool the tea down.

EVALUATE Despite your best effort, the tea will be watery. Putting the ice in a bag will prevent the melted ice water from mixing with the tea. More importantly, we see how we must proceed stepwise through calorimetry problems.

5: Heat flow through three bars

A composite rod is made up of three equal lengths and cross sections of aluminum, brass, and copper. The free aluminum end is maintained at 100°C, and the free end of the copper rod is maintained at 0°C. If the surface of the rod is insulated to prevent radial heat flow, find the temperature at each junction.

Solution

IDENTIFY We will use the heat current through the rods to find the temperatures at the rod junctions—the target variables.

SET UP A sketch of the rod is shown in Figure 17.2. The heat current is the same through each segment of the rod, so we'll write the heat equations for each segment and set them equal to each other to solve the problem.

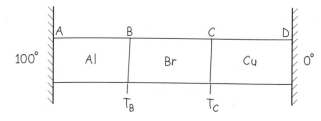

Figure 17.2 Problem 5 sketch.

EXECUTE The heat current through any rod is

$$H = kA\frac{T_H - T_C}{L}.$$

We can write the heat current per unit area times the length through the individual rods as

$$\frac{H_{AB}L}{A} = k_{Al}(T_H - T_C) = (205 \text{ W/m} \cdot \text{K})(100°C - T_B),$$

$$\frac{H_{BC}L}{A} = k_{Br}(T_H - T_C) = (109 \text{ W/m} \cdot \text{K})(T_B - T_C),$$

$$\frac{H_{CD}L}{A} = k_{Co}(T_H - T_C) = (385 \text{ W/m} \cdot \text{K})(T_C - 0°C).$$

These expressions must all be equal, since they each reference the same heat current, area, and length of the rod segment. Setting the last two equations equal to each other results in

$$(109 \text{ W/m} \cdot \text{K})T_B - (109 \text{ W/m} \cdot \text{K})T_C = (385 \text{ W/m} \cdot \text{K})T_C,$$

or

$$T_B = \frac{385 + 109}{109}T_C = 4.53T_C.$$

Setting the first and last equations equal to each other gives

$$(205 \text{ W/m} \cdot \text{K})(100°C) - (205 \text{ W/m} \cdot \text{K})T_B = (385 \text{ W/m} \cdot \text{K})T_C.$$

Replacing T_B yields

$$205(100°C) - 205(4.53T_C) = 385\,T_C,$$
$$T_C = 15.6°C.$$

Then

$$T_B = 70.7°C.$$

The aluminum-brass junction is at 70.7°C and the brass/copper junction is at 15.6°C.

EVALUATE Although one might expect the three equal segments to have equal temperature differences, we see that the varying thermal conductivities of the segments caused a nonuniform temperature distribution. The segment with the highest thermal conductivity (copper) had the smallest temperature difference between its ends, and the segment with the lowest thermal conductivity (brass) had the greatest temperature difference between its ends.

6: Time required to melt a block of ice

A long steel rod that is insulated to prevent heat loss along its sides is in perfect thermal contact with a large container of boiling water at one end and a 3.0-kg block of ice at the other. The steel rod is 1.2 m long with cross-sectional area 3.50 cm². How long does it take for the block of ice to melt? The ice block is initially at 0°C.

Solution

IDENTIFY We'll combine our knowledge of heat conduction with our knowledge of heat of fusion to solve this problem. The target variable is the time required for the ice to melt.

SET UP A sketch of the problem is shown in Figure 17.3. We begin by determining the heat required to melt the ice. We then find the rate of heat flow into the ice. With that information, we can find the time it takes for the ice to melt.

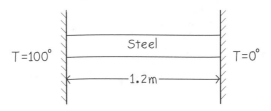

Figure 17.3 Problem 6 sketch.

EXECUTE The heat required to melt the ice is the heat of fusion for ice:

$$Q_{\text{melt}} = m_{\text{ice}}L_f = (3.0\ \text{kg})(3.34 \times 10^5\ \text{J/kg}) = 1.0 \times 10^6\ \text{J}.$$

The rate of heat flow is given by

$$H = \frac{\Delta Q}{\Delta t} = kA\frac{T_H - T_C}{L},$$

where k is the thermal conductivity, A and L are, respectively, the area and length of the bar, and T_H and T_C are, respectively, the temperatures of the hot and cold sides of the bar. We find that

$$H = \frac{\Delta Q}{\Delta t} = kA\frac{T_H - T_C}{L} = (50.2\ \text{W/(m}\cdot\text{K)})(6.5 \times 10^{-4}\ \text{m}^2)\frac{(100°C) - (0°C)}{(1.2\ \text{m})} = 2.7\ \text{W},$$

where we used 50.2 W/m/K as the thermal conductivity of steel. The time required to melt the ice is

$$\Delta t = \frac{Q_{\text{melt}}}{H} = \frac{(10^6\,\text{J})}{(2.7\,\text{W})} = 370,000\,\text{s}.$$

The time required to melt the ice is 370,000 s, or 103 hours.

EVALUATE We see that the thin steel bar is a relatively poor conductor of heat. Replacing the steel with a copper bar would increase the rate by almost a factor of 8, due to the differences in thermal conductivity. Increasing the rod's diameter and shortening the rod would also increase the rate of melting.

Try It Yourself!

1: Niagara Falls

Water flowing at a speed of 5.0-m/s-falls over a 50-m-high waterfall into a still pool below. Calculate the approximate rise in water temperature due to the conversion of mechanical energy into thermal energy

Solution Checkpoints

IDENTIFY AND SET UP Use energy conservation, equating the loss in mechanical energy to heat.

EXECUTE The water has kinetic energy and gravitational energy that together convert to heat. Equating the two forms of energy

$$\tfrac{1}{2}mv^2 + mgh = mc\Delta T.$$

The rise in temperature is 0.12°C.

EVALUATE Did you need to know the mass of the water?

2: Melting ice, again

A copper calorimeter of mass 2.0 kg initially contains 1.5 kg of ice at $-10°C$. (a) What will the final temperature be if the heat added is 5×10^5 J? (b) What will the final temperature be if the heat added is 10^6 J?

Solution Checkpoints

IDENTIFY AND SET UP Use heat capacity and heat of fusion to determine how the temperature of the ice increases with the heat provided. Follow a series of steps and determine whether the heat at each step exceeds the heat provided.

EXECUTE (a) Examining Problem 4, we see that 5×10^5 J would be exhausted during the melting phase of the problem. The final temperature is 0°C.

   (b) With 10^6 J of heat, all of the ice melts but the temperature doesn't reach 100°C. The temperature can be found by adding up the heat required in each step:

$$\Delta Q = m_{\text{ice}}c_{\text{ice}}(10°C) + m_{\text{copper}}c_{\text{copper}}\Delta T + m_{\text{ice}}c_{\text{water}}(\Delta T - 10°C) + m_{\text{ice}}L_f.$$

The final temperature is 64.8°C.

EVALUATE Why is $-10°C$ required in the term expressing the heat capacity of water?

18

Thermal Properties of Matter

Summary

In this chapter, we extend our investigation into thermodynamics, viewing systems from both the macroscopic and microscopic perspectives and building links between the two perspectives. We will learn about equations of state for materials and examine the ideal-gas equation as one such equation. This investigation will allow us to build a model for the kinetic energy of individual molecules and predict the behavior of gases. We will define thermodynamic systems and examine the energy of those systems. This analysis will lead to the first law of thermodynamics and thermodynamic processes. Four common thermodynamics processes will be highlighted, and their implications for ideal gases examined.

Objectives

After studying this chapter, you will understand

- How to define the mole and Avogadro's number.
- How to define equations of state and how to apply the ideal-gas equation.
- How to determine the kinetic energy of gases and how to apply that energy to individual particles.
- The origins of molar heat capacities for materials and gases.
- How to apply the first law of thermodynamics.
- The four common thermodynamic processes and how to apply them to find the changes in heat, work, and internal energy of thermodynamic systems.

Concepts and Equations

Term	Description
Mole	One mole (mol) is the amount of substance that contains the same number of elementary units as there are atoms in 0.012 kg of carbon 12. The number of molecules in a mole is Avogadro's number, $N_A = 6.022 \times 10^{23}$ molecules per mole. The molar mass is the mass of 1 mole of a substance.
Equation of State	An equation of state expresses the relation among pressure, temperature, and volume of a certain amount of a substance in equilibrium. The pressure p, volume V, and absolute temperature T are the state variables.
Ideal-Gas Equation	The ideal-gas equation is the equation of state for an ideal gas that approximates the behavior of a real gas at a low pressure and a high temperature. The pressure p, temperature T, volume V, and number of moles, n, of the gas are related by $$pV = nRT,$$ where R is the ideal-gas constant. In SI units, when pressure is given in Pa and volume is given in m^3, $R = 8.3145 \text{ J}/(\text{mol} \cdot \text{K})$.
Kinetic Theory of Gases	The total translational kinetic energy K_{tr} of all the molecules in an ideal gas is proportional to the temperature T and quantity of gas, n, in moles. Expressed as an equation, the total translational kinetic energy is $$K_{tr} = \tfrac{3}{2}nRT.$$ For a single molecule, the average translational kinetic energy is $$K_{av} = \tfrac{3}{2}kT,$$ where $k = R/N_A = 1.381 \times 10^{-23} \text{ J}/(\text{molecule} \cdot \text{K})$ is the Boltzmann constant. The mean free path of molecules in an ideal gas is given by $$\lambda = vt_{mean} = \frac{V}{4\pi\sqrt{2}r^2N}.$$
Molar Heat Capacity	The amount of heat Q needed for a temperature change ΔT is $$Q = nC\Delta T,$$ where n is the number of moles of the substance and C is the molar heat capacity. The molar heat capacity at constant volume is given in certain cases by $$C_V = \frac{3}{2}R \quad (\text{monatomic gas}),$$ $$C_V = \frac{5}{2}R \quad (\text{diatomic gas}),$$ $$C_V = 3R \quad (\text{monatomic solid}).$$
Molecular Speeds	The speeds of molecules in an ideal gas are given by the Maxwell–Boltzmann distribution: $$f(v) = 4\pi\left(\frac{m}{2\pi kT}\right)^{3/2}v^2 e^{-mv^2/2kT}.$$
	The quantity $f(v)\,dv$ describes the fraction of molecules with speeds between v and $v + dv$.
Phases of Matter	Ordinary matter exists in solid, liquid, and gas phases. A phase diagram shows the conditions under which two phases can coexist in phase equilibrium. All three phases can coexist at the triple point.

Conceptual Questions

1: Don't hold your breath

Explain why scuba divers are taught not to hold their breath as they ascend to the surface from depths under the water.

Solution

IDENTIFY, SET UP, AND EXECUTE We know from fluid statics that pressure increases with depth in water. The ideal-gas equation states that pressure and volume are inversely proportional for a given temperature and quantity of gas. Ascending to the surface reduces the ambient pressure, causing an increase in volume. (We assume that the temperature is constant in the water). By holding her breath, a scuba diver traps a quantity of air inside her lungs. As the pressure decreases upon her ascent, her lungs expand, possibly damaging some lung tissue. If the diver exhales during the ascent, the pressure cannot build to dangerous levels.

EVALUATE This problem combines our knowledge of fluid statics and our knowledge of ideal gases and helps illustrate the relation between pressure and volume. High-altitude weather balloons also expand as they rise, so they are partially filled at the ground to prevent the balloons from bursting as they ascend.

2: Atmosphere on the earth and moon

Why does the earth, but not the moon, have an atmosphere?

Solution

IDENTIFY, SET UP, AND EXECUTE The escape velocity of molecules on the earth is about 11 km/s, much higher than the average rms speed of molecules in the atmosphere. Without sufficient speed, the molecules remain near earth, thus creating an atmosphere.

 The gravitational potential on the moon's surface is about 20 times weaker than that on the earth's surface, so the escape speed is about 20 times less, or 2400 m/s. This lesser escape speed greatly enhances the probability that molecules of whatever atmosphere the moon might have had have all escaped into space.

EVALUATE The problem illustrates how molecular motion can help explain common physics phenomena.

Problems

1: Changing volume in a diving bell

A diving bell (a circular cylinder 3.0 m high, open at the bottom) is lowered into a lake. By how much does the water rise as the bell is lowered 75 m? The surface temperature of the lake is 25°C and the temperature at the 75 m depth is 15°C.

Solution

IDENTIFY We assume that the gas is ideal, so we use the ideal-gas equation to relate the surface values of pressure, temperature, and volume to the values at depth. The target value is the height of the water in the diving bell at depth.

SET UP Figure 18.1 shows a diagram of the situation. We'll use fluid statics to relate the pressure at the surface to the pressure at depth. These two relations will be combined to find the final height of water in the bell.

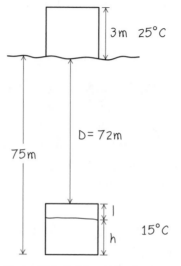

Figure 18.1 Problem 1 sketch.

EXECUTE The same amount of gas is trapped inside the bell both at the surface and at depth; therefore,

$$\frac{p_S V_S}{T_S} = \frac{p_D V_D}{T_D} = \text{constant},$$

where the subscript S indicates the value at the surface and subscript D indicates the value at depth. Substituting the known values gives

$$\frac{(1.01 \times 10^5 \text{ Pa})(A)(3.0 \text{ m})}{(298 \text{ K})} = \frac{p_D A l}{(288 \text{ K})},$$

where we replaced the volume of the cylinder with its area times its height. There two unknowns in this equation:

$$p_D l = (288 \text{ K})\frac{(1.01 \times 10^5 \text{ Pa})(3.0 \text{ m})}{(298 \text{ K})} = 2.93 \times 10^5 \text{ Pa m}.$$

We can find the pressure at depth from fluid statics, using

$$p_D = p_S + \rho g(D + l).$$

Substituting the known values yields

$$p_D = (1.01 \times 10^5 \text{ Pa}) + (1 \times 10^3 \text{ kg/m}^3)(9.8 \text{ m/s}^2)(72 \text{ m} + l),$$
$$p_D = (8.07 \times 10^5 \text{ Pa}) + (7.06 \times 10^5 \text{ Pa/m})l.$$

This equation also has two unknowns. Combining the two equations to eliminate p_D results in

$$p_D = \frac{(2.93 \times 10^5 \text{ Pa m})}{l} = (8.07 \times 10^5 \text{ Pa}) + (7.06 \times 10^5 \text{ Pa/m})l.$$

Rearranging terms gives

$$(7.06/\text{m}^2)l^2 + (8.07/\text{m})l - 2.93 = 0.$$

This is a quadratic equation with solutions $l = 0.290, -1.43$ m. The negative root is nonphysical, so the correct l is 0.29 m. The water rises 3.0 m $-$ 0.29 m, or 2.7 m, as the bell descends.

EVALUATE This problem illustrates how the increased pressure at depth reduces the volume of gas in the diving bell. If you consider the reverse process, you can see how the volume would increase as the bell rises to the surface, as we discussed in Conceptual Question 1. You can also try both situations by using a bucket of water. Submerge an inverted glass in a bucket of water, and see how the water level in the glass rises as the glass is lowered. Then use a hose to add air to the bottom of an inverted glass at the bottom of the bucket. As you raise the glass, you should see air leaving it.

2: Spacing between hydrogen molecules

Find the average spacing between H_2 molecules, assuming that the molecules are at the vertices of a fictitious cubic structure, in (a) gaseous H_2 at STP, (b) liquid H_2 at 20 K, where the density is 41,060 mol/m^3, and (c) solid H_2 at 4.2 K, where the molar volume is 22.91×10^{-6} m^3/mol.

Solution

IDENTIFY We will use the definitions of density, Avogadro's number, and molar mass to find the average spacing.

SET UP We'll assign a spacing of a to the distance between H_2 molecules, giving a volume of a^3 per molecule. One mole of gas occupies 22.4 L at STP, and 1 L is 0.001 m^3.

EXECUTE (a) For the gaseous H_2, we multiply the volume of one molecule by N_A to get the volume at STP:

$$N_A a^3 = 22.4 \text{ L/mol} = 22.4 \times 10^{-3} \text{ m}^3/\text{mol}.$$

Solving for a gives

$$a = \sqrt[3]{\frac{22.4 \times 10^{-3} \text{ m}^3/\text{mol}}{N_A}} = \sqrt[3]{\frac{22.4 \times 10^{-3} \text{ m}^3/\text{mol}}{6.203 \times 10^{23}}} = 3.34 \times 10^{-9} \text{ m}.$$

(b) We are given that each cubic meter contains 41,060 moles of liquid H_2. In equation form, this is

$$1 \text{ m}^3 = 41,060 \, N_A a^3.$$

Solving for a results in

$$a = \sqrt[3]{\frac{1 \text{ m}^3}{41,060 \, N_A}} = \sqrt[3]{\frac{1 \text{ m}^3}{(41,060)(6.203 \times 10^{23})}} = 3.44 \times 10^{-10} \text{ m}.$$

(c) We are given the molar volume:

$$N_A a^3 = 22.91 \times 10^{-6} \text{ m}^3/\text{mol}.$$

Solving for a produces

$$a = \sqrt[3]{\frac{22.91 \times 10^{-6} \text{ m}^3/\text{mol}}{N_A}} = \sqrt[3]{\frac{22.91 \times 10^{-6} \text{ m}^3/\text{mol}}{6.203 \times 10^{23}}} = 3.37 \times 10^{-10} \text{ m}.$$

The average spacing is 3.34×10^{-9} m for the gaseous H_2, 3.44×10^{-10} m for the liquid H_2, and 3.37×10^{-10} m for the solid H_2.

EVALUATE We see that the average spacing for the liquid and solid H_2 is similar and about 10 times closer than the spacing for the gaseous H_2.

3: Spacing in a vacuum

A common type of laboratory vacuum pump produces an ultimate pressure of 10 microns. How many molecules (in moles) of gas are present in a volume of 0.15 m^3 reduced to that pressure at 300 K?

Solution

IDENTIFY We will use the ideal-gas law to solve for the number of molecules in the volume—the target variable.

SET UP To use the ideal-gas law, we need to know the pressure in standard units. One micron is a measure of pressure in terms of the height of mercury. One atmosphere is 760 mm of mercury; one micron is one-thousandth of a millimeter of mercury.

EXECUTE First convert the pressure to atmospheres, using the information provided:

$$10 \text{ microns} = (10 \times 10^{-6})\left(\frac{1 \text{ atm}}{0.760 \text{ m}}\right) = 1.32 \times 10^{-5} \text{ atm}.$$

The ideal-gas law is

$$pV = nRT.$$

Solving for n and using the molar gas constant yields

$$n = \frac{pV}{RT} = \frac{(1.32 \times 10^{-5} \text{ atm})(1.5 \times 10^2 \text{ m}^3)}{(0.08206 \text{ L} \cdot \text{atm/mol} \cdot \text{K})(300 \text{ K})} = 8.04 \times 10^{-5} \text{ mol}.$$

There is 8.04×10^{-5} of a mole of molecules in the container after evacuating.

EVALUATE This may seem like a small number, but there are almost 5×10^{19} molecules remaining in the volume. One of the toughest challenges in physics research is creating ultrahigh vacuums to study the behavior of small numbers of particles. Physicists don't want collisions with remaining air molecules to interfere with the molecules whose behavior they are studying.

Did we assume that the gas was an ideal gas? Yes, we took the gas to be an ideal gas. This assumption is valid because the gas is at low pressure.

4: Mixing gases

A 1-liter flask at 293 K contains a mixture of 3 g of N_2 and 3 g of H_2 gas. Assuming that the gases behave as ideal gases, (a) calculate the partial pressure exerted by both gases and (b) calculate the rms speeds of the two gases.

Solution

IDENTIFY We will use the ideal-gas law and kinetic theory to solve the problem. The target variables are the partial pressures and rms speeds of the two gases.

SET UP Partial pressure is the pressure exerted by each gas separately. We'll use the ideal-gas law to find the partial pressure of each gas. The rms speed may be calculated by an expression found in the text.

EXECUTE (a) The volume and temperature are given, so we need to convert the amount of each gas to moles. We have

$$n_{N_2} = \frac{3\text{ g}}{28\text{ g/mol}} = \frac{3}{28}\text{ mol},$$

$$n_{H_2} = \frac{3\text{ g}}{2\text{ g/mol}} = \frac{3}{2}\text{ mol}.$$

The ideal-gas law is

$$pV = nRT.$$

Solving for p for each gas yields

$$p_{N_2} = \frac{nRT}{V} = \frac{\left(\frac{3}{28}\text{ mol}\right)(8.314\text{ J/mol}\cdot\text{K})(293\text{ K})}{10^{-3}\text{ m}^3} = 2.61 \times 10^5\text{ Pa} = 2.6\text{ atm},$$

$$p_{H_2} = \frac{nRT}{V} = \frac{\left(\frac{3}{2}\text{ mol}\right)(8.314\text{ J/mol}\cdot\text{K})(293\text{ K})}{10^{-3}\text{ m}^3} = 3.65 \times 10^6\text{ Pa} = 36\text{ atm}.$$

(b) The rms speed is found from

$$v_{\text{rms}} = \sqrt{\frac{3RT}{M}}.$$

Solving for our gases gives

$$v_{N_2} = \sqrt{\frac{3RT}{M}} = \sqrt{\frac{3(8.314\text{ J/mol}\cdot\text{K})(293\text{ K})}{(28 \times 10^{-3}\text{ kg/mol})}} = 511\text{ m/s},$$

$$v_{H_2} = \sqrt{\frac{3RT}{M}} = \sqrt{\frac{3(8.314\text{ J/mol}\cdot\text{K})(293\text{ K})}{(2 \times 10^{-3}\text{ kg/mol})}} = 1910\text{ m/s}.$$

EVALUATE We see that both the pressure and the rms speed are higher for the H_2 gas, consistent with its smaller mass.

5: Escape velocities of the sun and earth

Calculate the escape velocities at the surface of the sun and the earth, and determine whether any molecules have rms thermal speeds comparable with these escape velocities at 300 K.

Solution

IDENTIFY We will use energy conservation to determine the escape velocities and kinetic theory to find the rms speeds. The target variables are the escape velocities at the surface of the sun and earth and the rms speeds for gases at 300 K. The mass and radius values are found in Appendix F of the text.

SET UP For a molecule to escape from the sun or the earth, all of the molecule's gravitational potential energy must convert to kinetic energy. We use that relation to solve for the minimum escape velocity. The rms speed is given by an expression found in the text.

EXECUTE The escape velocity is found from energy conservation:

$$\frac{GMm}{R} = \tfrac{1}{2}mv^2.$$

Solving for the velocity gives

$$v = \sqrt{\frac{2GM}{R}}.$$

Solving for the escape velocities of the sun and the earth yields

$$v_{sun} = \sqrt{\frac{2GM_{sun}}{R_{sun}}} = \sqrt{\frac{2(6.67 \times 10^{-11}\,\text{N} \cdot \text{m}^2/\text{kg}^2)(2 \times 10^{30}\,\text{kg})}{7.0 \times 10^{8}\text{m}}} = 6.2 \times 10^{5}\,\text{m/s},$$

$$v_{earth} = \sqrt{\frac{2GM_{earth}}{R_{earth}}} = \sqrt{\frac{2(6.67 \times 10^{-11}\,\text{N} \cdot \text{m}^2/\text{kg}^2)(6.0 \times 10^{24}\,\text{kg})}{6.38 \times 10^{6}\text{m}}} = 1.12 \times 10^{4}\,\text{m/s}.$$

The rms speed is found from

$$v_{rms} = \sqrt{\frac{3RT}{M}}.$$

The hydrogen atom has the highest rms speed of any gaseous atom, so we calculate the speed for hydrogen. For a hydrogen atom at 300 K,

$$v_{H} = \sqrt{\frac{3RT}{M}} = \sqrt{\frac{3(8.314\,\text{J/mol} \cdot \text{K})(300\,\text{K})}{(1 \times 10^{-3}\,\text{kg/mol})}} = 2700\,\text{m/s}.$$

The rms speed is 2700 m/s, less than the escape velocity of the sun or the earth.

EVALUATE We see that it is unlikely for any gas to have sufficient rms speed to escape the sun or the earth. The sun's average surface temperature is roughly 6000 K, corresponding to 12,000 m/s, a speed still too low for escape. Why does the moon not have an atmosphere?

Practice Problem: Find the escape velocity of a particle on the moon. *Answer*: 2370 m/s, enough for most gases to escape.

Try It Yourself!

1: Helium gas

Helium gas is admitted to a volume of 200 cm^3 at a temperature of 77 K until the pressure is equal to 1 atm. (a) If the temperature of the container is raised to 20°C, what will the pressure inside the container be? (b) If the system has a relief valve that will not permit the pressure to exceed 1 atm, what fraction of gas remains at 20°C?

Solution Checkpoints

IDENTIFY AND SET UP Use the ideal-gas law to solve the problem. Is that law a valid approximation?

EXECUTE (a) For the closed system, of the number of moles, pressure, temperature, and volume, which changes as the temperature rises? Only pressure and temperature change, so their ratio remains constant:

$$\frac{p_1}{T_1} = \frac{p_2}{T_2}.$$

The final pressure is 3.8 atm.

(b) For the valved system, what remains constant? Both the moles and temperature change, yielding

$$n_1RT_1 = n_2RT_2.$$

26.3% of the gas remains.

EVALUATE How could the system be changed so that both the pressure and the number of moles remain constant?

2: Nitrogen gas

(a) Calculate the volume occupied by 1 mole of nitrogen gas at the critical temperature (162.2 K) and critical pressure $(33.9 \times 10^5 \, \text{N/m}^2)$. Express your answer as a ratio of the volume to the known critical volume $(90.1 \times 10^{-6} \, \text{m}^3)$. (b) Calculate the pressure at the critical volume and temperature.

Solution Checkpoints

IDENTIFY AND SET UP Treat the nitrogen as an ideal gas and use the ideal-gas law to solve the problem.

EXECUTE (a) The ideal-gas law can be used to find the volume:

$$V = \frac{nRT}{p}.$$

The volume is $3.44V_C$.
(b) The critical pressure is $3.44p_c$.

EVALUATE What do the critical pressure, temperature, and volume refer to? Do you think the results are valid?

3: RMS speed values

In a gas at 300 K that is a mixture of the diatomic molecules H_2 (2 g/mol) and D_2 (4 g/mol), find the rms speeds of the molecules

Solution Checkpoints

IDENTIFY AND SET UP Kinetic theory gives the rms speed of molecules.

EXECUTE The rms speed is found from

$$v = \sqrt{\frac{3RT}{M}}.$$

The rms speed is 1930 m/s for hydrogen and 1370 m/s for deuterium (D_2).

EVALUATE Could you devise a method for separating these two molecules by using the differences in their rms speeds? Explain.

The First Law of Thermodynamics

Summary

In this chapter, we investigate and quantify thermodynamic processes—processes that exchange heat and do work. We will examine thermodynamic systems and energy in these systems. Our examination will lead us to the first law of thermodynamics and thermodynamic processes. Four common thermodynamics processes will be highlighted, and the implications of those processes for ideal gases will be examined.

Objectives

After studying this chapter, you will understand

- How to define thermodynamic processes.
- How heat is transferred and work is done in a thermodynamic process.
- The definition of, and how to apply, the first law of thermodynamics.
- How a path between initial and final states affects a thermodynamic process.
- The four common thermodynamic processes (adiabatic, isochoric, isobaric, and isothermal).
- How to apply the common thermodynamic processes to find the changes in heat, work, and internal energy of thermodynamic systems.
- How ideal gases are described in the common thermodynamic processes.

Concepts and Equations

Term	Description
Heat and Work in Thermodynamic Processes	A thermodynamic system may exchange energy with its surroundings by heat transfer or by mechanical work. The work done by the system is given by $$W = \int_{V_1}^{V_2} p\,dV$$ $$= p(V_2 - V_1) \quad \text{(constant pressure only)}.$$ In any thermodynamic process, the heat added to the system and the work done by the system depend on the steps the system takes from its initial to its final states, as well as on the initial and final states themselves.
First Law of Thermodynamics	The first law of thermodynamics states that when heat Q is added to a system while work W is performed by the system, the internal energy U changes by $$\Delta U = Q - W.$$ For infinitesimal changes, $$dU = dQ - dW.$$ The internal energy of any thermodynamic system depends only on its state. The change in internal energy in any process depends only on the initial and final states.
Thermodynamic Processes	Common thermodynamic processes include the adiabatic process, in which no heat flows into or out of the system $(Q = 0)$; the isochoric process, in which the volume remains constant $(W = 0)$; the isobaric process, in which the pressure remains constant $[W = p(V_2 - V_1)]$; and the isothermal process, in which the temperature remains constant.
Properties of an Ideal Gas	The internal energy of an ideal gas depends only on its temperature, not its pressure or volume. The molar heat capacity at constant volume (C_V) and the molar heat capacity at constant pressure (C_p) for an ideal gas are related by $$C_p = C_V + R.$$ For an adiabatic process in an ideal gas, both $TV^{\gamma-1}$ and pV^{γ} are constant, where $\gamma = C_p/C_V$. The work done by an ideal gas during an adiabatic expansion is given by $$W = nC_V(T_1 - T_2)$$ $$= \frac{C_V}{R}(p_1V_1 - p_2V_2)$$ $$= \frac{1}{\gamma - 1}(p_1V_1 - p_2V_2).$$

Conceptual Questions

1: *pV* diagrams

One mole of helium gas is placed in a sealed container and undergoes an isochoric process that results in a doubling of the helium's pressure. Next, the gas undergoes an adiabatic process until the volume of the container is tripled. It then undergoes an isobaric expansion which results in a volume that is four times its original volume. Finally, the helium undergoes an isothermal compression that leaves the container with the same volume that it had after the first process. Sketch a *pV* diagram for this combined process.

Solution

SET UP AND SOLVE Figure 19.1 shows the resulting *pV* diagram. The diagram starts at point *a* with initial pressure p_0 and volume V_0. Then comes the isochoric process, at constant volume, represented by a vertical line to point *b,* where the pressure has doubled. Next is the adiabatic process, in which no heat is exchanged. This process follows the path to point *c,* where the volume has tripled. Next is the isobaric process, carried out at constant pressure and represented by a horizontal line to point *d,* where the volume has increased by V_0 from point *c.* Finally, in the isothermic process, the segment follows the path to point *e,* where the pressure has increased to $2p_0$.

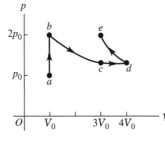

Figure 19.1 Question 1.

REFLECT This problem helps clarify the differences in the four common thermodynamic processes. *pV* diagrams are a valuable aid in solving problems involving these processes. The diagrams will help lead us through a particular problem, as well as provide a check on our results.

2: Internal energy in a thermodynamic process

A container of argon gas undergoes a multistep process. First, it undergoes an isobaric expansion that triples its volume. Next, it goes through an isochoric process that results in a doubling of the argon's pressure. Then it cools adiabatically by 50 K. After that, it undergoes a second isochoric process that doubles its volume. Finally, it undergoes isobaric compression that leaves it at its initial temperature. Find the total change in internal energy.

Solution

SET UP AND SOLVE The change in the internal energy of an ideal gas depends only upon temperature. Argon is an ideal gas. Since the final temperature is the same as the initial temperature, the total change in internal energy is zero.

REFLECT This complicated scenario shows that we need to focus on the important parts of the process to interpret the results properly. Although it is trivial to find the change in internal energy, it would be cumbersome to find the heat added to the system.

Problems

1: Adiabatic compression of helium

Helium gas is expanded adiabatically from a 12-liter volume at STP to a 33-liter volume. Find the final temperature and pressure of the gas and the work done on the gas.

Solution

IDENTIFY Since the helium is expanded adiabatically, both pV^γ and $TV^{\gamma-1}$ are constant during the process. The target variables are the final temperature and pressure of the gas and the work done on the gas.

SET UP For helium, $\gamma = 1.67$ (from Table 19.1 in the text). *Standard temperature and pressure* (STP) refers to a temperature of 273 K and a pressure of 1 atmosphere.

EXECUTE To find the final pressure, we use the relation

$$pV^\gamma = \text{constant} = p_1 V_1^\gamma = p_2 V_2^\gamma,$$

where the subscripts 1 and 2 indicate "before" and "after," respectively. Rearranging terms to find the final pressure, we obtain

$$p_2 = \frac{p_1 V_1^\gamma}{V_2^\gamma} = \frac{(1.01 \times 10^5 \text{ Pa})(12 \,\ell)^{1.67}}{(33 \,\ell)^{1.67}} = 1.86 \times 10^4 \text{ Pa}.$$

To find the final temperature, we use the relation

$$TV^{\gamma-1} = \text{constant} = T_1 V_1^{\gamma-1} = T_2 V_2^{\gamma-1}.$$

Rearranging terms in the preceding equation gives a final temperature of

$$T_2 = \frac{T_1 V_1^{\gamma-1}}{V_2^{\gamma-1}} = \frac{(273 \text{ K})(12 \,\ell)^{0.67}}{(33 \,\ell)^{0.67}} = 139 \text{ K}.$$

The work done by an ideal gas in an adiabatic process is

$$W = nC_V(T_1 - T_2).$$

For helium, C_V is 12.47, from Table 19.1. The number of moles is

$$n = \frac{12 \text{ L}}{22.4 \text{ L}} = 0.536 \text{ mol}.$$

The work done by the gas is

$$W = nC_V(T_1 - T_2) = (0.536 \text{ mol})(12.47 \text{ J/mol} \cdot \text{K})(273 \text{ K} - 139 \text{ K}) = 895 \text{ J}.$$

The work done *on* the gas is the opposite, or −895 J. The final pressure is 1.86×10^4 Pa and the final temperature is 139 K.

EVALUATE We see that both the temperature and the pressure decreased in this adiabatic expansion. That makes sense, since no heat was transferred into or out of the system, so having a larger volume required a lower pressure and temperature.

Would you find the same final temperature if you used the ideal-gas equation? If you check, you'll find that you indeed do find the same final temperature.

2: Isochoric and isobaric process with helium

Two moles of helium gas are taken from point *a* to point *c* in the diagram shown in Figure 19.2. Find the change in internal energy along path *abc*.

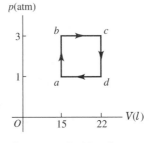

Figure 19.2 Problem 2.

Solution

IDENTIFY The target variable is the change in internal energy.

SET UP We break the process up into two segments, one from *a* to *b* and one from *b* to *c*. The first segment is an isochoric process (carried out at constant volume) and the second is an isobaric process (carried out at constant pressure). We'll use the relations for those segments to determine the work, heat, and temperature changes. We'll combine the work and heat changes to find the change in internal energy.

EXECUTE We find the change in internal energy along *ab* by first finding the change in temperature along *ab*:

$$\Delta T_{ab} = \frac{(p_b - p_a)V_a}{nR} = \frac{(3.03 \times 10^5 \text{ Pa} - 1.01 \times 10^5 \text{ Pa})(15 \ \ell)(10^{-3} \text{ m}^3/\ell)}{(2 \text{ mol})(8.31 \text{ J/mol/K})} = 182 \text{ K}.$$

The heat transferred during *ab* is then

$$Q_{ab} = nC_V\Delta T_{ab} = (2 \text{ mol})(12.47 \text{ J/mol/K})(182 \text{ K}) = 4500 \text{ J},$$

where we used the molar heat capacity at constant volume for helium ($C_V = 12.47$ J/mol/K, from Table 19.1). The work done during segment *ab* is zero, since that segment is isochoric. The change in internal energy for segment *ab* is then

$$\Delta U_{ab} = Q_{ab} - W_{ab} = 4500 \text{ J} - 0 = 4500 \text{ J}.$$

For segment *bc*, we follow the same procedure. The change in temperature along *bc* is

$$\Delta T_{bc} = \frac{p_b(V_c - V_b)}{nR} = \frac{(3.03 \times 10^5 \text{ Pa})(2.2 \times 10^{-2} \text{ m}^3 - 1.5 \times 10^{-2} \text{ m}^3)}{(2 \text{ mol})(8.31 \text{ J/mol/K})} = 127 \text{ K}.$$

The heat transferred during *bc* is

$$Q_{bc} = nC_P\Delta T_{bc} = (2 \text{ mol})(20.78 \text{ J/mol/K})(127 \text{ K}) = 5300 \text{ J},$$

where we used the molar heat capacity at constant pressure for helium ($C_P = 20.78$ J/mol/K, from Table 19.1). The work done during segment *bc* is

$$W_{bc} = p_b\Delta V_{bc} = (3.03 \times 10^5 \text{ Pa})((2.2 \times 10^{-3} \text{ m}^3) - (1.5 \times 10^{-3} \text{ m}^3)) = 2100 \text{ J}.$$

The change in internal energy during segment bc is

$$\Delta U_{bc} = Q_{bc} - W_{bc} = 5300 \text{ J} - 2100 \text{ J} = 3200 \text{ J}.$$

The total change in internal energy is

$$\Delta U = \Delta U_{ab} + \Delta U_{bc} = 4500 \text{ J} + 3200 \text{ J} = 7700 \text{ J}.$$

The total change in internal energy in the system is 7700 J.

EVALUATE We see that by breaking up a process into segments, determining the type of process that takes place during each segment, and knowing which variables change and which remain constant during each segment, one can easily find the change in internal energy.

How should the change in internal energy along path adc compare with the change along path abc? They should be the same for helium, an ideal gas.

Practice Problem: Find the change in internal energy along segments ad and dc, and compare their sums with the change in internal energy along path abc. *Answer:* $\Delta U_{ad} = 1100$ J, $\Delta U_{dc} = 6600$ J, $\Delta U_{adc} = 7700$ J.

3: Monatomic gas process

One mole of an ideal monatomic gas starts at point A in Figure 19.3 ($T = 273$ K, $p = 1$ atm) and undergoes an adiabatic expansion to point B, where the volume of the gas has doubled. These two processes are followed by an isothermal compression to the original volume at point C and an isobaric increase to the original point A. Find (a) the temperature at point B, (b) the pressure at point C, and (c) the total work done for the entire cycle. Take γ to be 5/3.

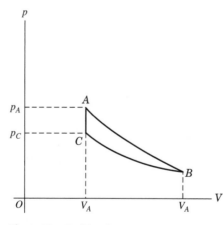

Figure 19.3 Problem 3.

Solution

IDENTIFY The target variables are the temperature at point B, the pressure at point C, and the total work done during the process.

SET UP We'll break the process up into the three segments shown in the figure. We'll use the relations for those segments to determine the temperature, pressure, and work.

EXECUTE (a) For the adiabatic process, we use the relation

$$TV^{\gamma-1} = \text{constant} = T_1 V_1^{\gamma-1} = T_2 V_2^{\gamma-1}.$$

Rearranging terms to find the temperature at point B gives

$$T_2 = \frac{T_1 V_1^{\gamma-1}}{V_2^{\gamma-1}} = \frac{(273 \text{ K})(V)^{0.67}}{(2V)^{0.67}} = 172 \text{ K}.$$

(b) The temperature at point C is the same as at point B, since the second step is isothermal. Also, the volume is the same at point C as it was at point A. We use the standard ideal-gas law:

$$\frac{p_1 V_1}{T_1} = \frac{p_3 V_3}{T_3}.$$

Solving for p_3 gives

$$p_3 = p_1 \frac{T_3}{T_1} = (1 \text{ atm})\frac{(172 \text{ K})}{(273 \text{ K})} = 0.630 \text{ atm}.$$

(c) The total work done by the gas during the entire cycle is the sum of the separate amounts of work done during each cycle. The path AB is adiabatic, so no heat is exchanged and the work is

$$W_{AB} = -C_V(T_2 - T_1) = -\left(\tfrac{3}{2}(8.314 \text{ J/mol} \cdot \text{K})\right)(172 \text{ K} - 273 \text{ K}) = 1260 \text{ J}.$$

For segment BC, the temperature is constant. The work is given by the integral

$$W = \int_{V_1}^{V_2} p \, dV$$

$$= \int_{V_2}^{V_3} \frac{RT}{V} dV = RT \ln V \Big|_{V_2}^{V_3}$$

$$= RT \ln(V_3/V_2) = (8.314 \text{ J/mol} \cdot \text{K})(172 \text{ K})\ln(V/2V)$$

$$= -991 \text{ J}.$$

No work is done during segment CA, since the system changes isobarically. The total work done is the sum, 268 J.

EVALUATE We solved this problem by breaking the process into segments and working through each segment to find the final state variables. Can you use the area inside the cycle to find the amount of work? The area is greater than zero, indicating positive work.

4: Expansion process for argon

One mole of argon is initially at 25°C and occupies a volume of 35 liters. The argon is first expanded at constant pressure until the volume is doubled and then expanded adiabatically until the temperature returns to 25°C. Find the total change in internal energy, the total work done by the argon, and the final volume and pressure of the argon.

Solution

IDENTIFY Sketch the process and then use the definitions to solve. The target variables are the total change in internal energy, the total work done, and the final volume and pressure.

SET UP Figure 19.4 shows the pV diagram for the process. We'll break the process up into two segments, one from a to b and one from b to c. The first segment is an isobaric process (carried out under constant pressure) and the second is adiabatic (no heat exchanged). We'll use the relations for those segments to determine the work, heat, and temperature changes. We'll combine these results to find the quantities of interest.

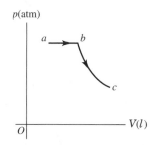

Figure 19.4 Problem 4 sketch.

EXECUTE The total change in internal energy is zero, since the internal energy of an ideal gas depends only on temperature and the final temperature is equal to the initial temperature.
Next, we find the temperature at point b. Segment ab is at constant pressure, so

$$\frac{V_a}{T_a} = \frac{V_b}{T_b}.$$

The temperature at b is

$$T_b = \frac{V_b}{V_a} T_a = \frac{(70\ \ell)}{(35\ \ell)}(273 + 25)\,\mathrm{K} = 596\ \mathrm{K}.$$

The heat supplied during ab is

$$Q_{ab} = nC_P\Delta T_{ab} = (1\ \mathrm{mol})(20.78\ \mathrm{J/mol/K})(596\ \mathrm{K} - 298\ \mathrm{K}) = 6190\ \mathrm{J},$$

where we used the molar heat capacity at constant pressure for argon ($C_P = 20.78\ \mathrm{J/mol/K}$, from Table 19.1). The heat transferred in bc is zero, since segment bc is adiabatic. The total heat supplied in the complete process is 6190 J. Because the total internal energy change is zero, the heat supplied must be equal to the work done by the argon. The work done by the argon is 6190 J.

Next, we find the final volume and pressure. To find the final volume in the adiabatic process (bc), we use the relation

$$TV^{\gamma-1} = \text{constant} = T_b V_b^{\gamma-1} = T_c V_c^{\gamma-1},$$

where $\gamma = 1.67$ for argon. Rearranging terms to find the final volume gives

$$V_c = \sqrt[\gamma-1]{\frac{T_b V_b^{\gamma-1}}{T_c}} = \sqrt[0.67]{\frac{(596\ \mathrm{K})(70\ \ell)^{0.67}}{(298\ \mathrm{K})}} = 197\ \mathrm{l}.$$

We can use the equation of state for an ideal gas to find the final pressure:

$$p_c = \frac{nRT_c}{V_c} = \frac{(1 \text{ mol})(8.31 \text{ J/mol/K})(298 \text{ K})}{(197 \times 10^{-3} \text{ m}^3)} = 1.26 \times 10^4 \text{ Pa}.$$

The final pressure is 12,600 Pa and the final volume is 197 liters.

EVALUATE We again see that we need to break the process up into segments and work through each segment to find the final state variables. We also see that the pV graph is useful in solving problems involving thermodynamic processes.

Try It Yourself!

1: Ideal-gas process

Consider n moles of an ideal gas that undergo the constant-volume and constant-pressure processes along the paths shown in Figure 19.5 from the initial state a to b, then from b to c, then from c to d, and then back from a to d. For each of these processes, calculate (a) the work done by the system and (b) the heat taken in by the system.

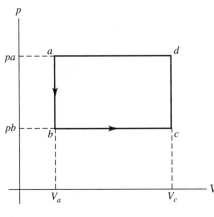

Figure 19.5 Try It Yourself 1.

Solution Checkpoints

IDENTIFY AND SET UP We break the process up into segments and use the relations for isochoric and isobaric processes.

EXECUTE (a) No work is done by the gas during the isochoric processes. For the isobaric processes, the work done by the gas is the pressure times the change in volume:

$$W_{b \to c} = p_b(V_c - V_a),$$
$$W_{d \to a} = p_a(V_a - V_c).$$

The work done from b to c is a positive quantity, and the work done from d to a is negative, resulting in negative total work. Work is done *on* the gas.

(b) The heat taken in by the ideal gas will be the number of moles, times the change in temperature, times the heat capacity at constant volume or constant pressure. This relationship gives

$$Q_{a \to b} = nC_V(T_b - T_a),$$

$$Q_{b \to c} = nC_p(T_c - T_b),$$

$$Q_{c \to d} = nC_V(T_d - T_c),$$

$$Q_{d \to a} = nC_p(T_a - T_d).$$

EVALUATE How does the heat taken in by the gas compare with the work done by the gas?

2: Ideal-gas process, version 2

A total of n moles of an ideal monatomic gas is taken from point 1 on the T_1 isotherm to point 3 on the T_2 isotherm along the path $1 \to 2 \to 3$ as shown in Figure 19.6. (a) Calculate the change in internal energy of the gas and the heat that must be added to it in this process. (b) Suppose the path $1 \to 4 \to 3$ is followed instead. Calculate the heat added along this path.

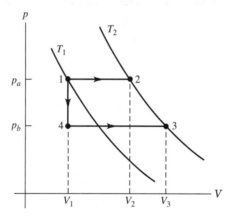

Figure 19.6 Try It Yourself 2.

Solution Checkpoints

IDENTIFY AND SET UP We break the process up into segments and use the relations for isothermic and isobaric processes.

EXECUTE (a) In going from 1 to 3, the change in internal energy depends on the temperature differences and not the path taken. This gives

$$\Delta U = nC_V(T_3 - T_1).$$

The heat taken in from 1 to 3 is

$$\Delta Q = nC_p(T_2 - T_1) + nRT \ln\left(\frac{V_3}{V_2}\right).$$

(b) The change in internal energy is the same as in (a). The heat is

$$\Delta Q = nC_V(T_2 - T_1) + p_3(V_3 - V_1).$$

EVALUATE How does the heat exchanged on the two paths differ. Why?

20 The Second Law of Thermodynamics

Summary

In this chapter, we will complete our investigation of thermodynamics, examining thermodynamic processes and the second law of thermodynamics. Heat engines and refrigerators transform heat into work or energy in cyclic processes. Thermal efficiency and performance coefficients for engines and refrigerators will be defined. The second law of thermodynamics limits the efficiency of engines and has profound implications in many physical processes. The second law can be quantified in terms of entropy, a measure of disorder. We will examine several common cyclic processes to aid our understanding of thermodynamics.

Objectives

After studying this chapter, you will understand

- How to define and identify reversible processes.
- How to analyze heat engines and refrigeration cycles.
- How to apply the second law of thermodynamics.
- How to calculate entropy for a variety of systems.
- How to apply thermodynamic principles to a variety of engine and refrigeration cycles.

Concepts and Equations

Term	Description
Directions of Thermodynamic Processes	Heat flows spontaneously from hotter objects to cooler objects in thermodynamic processes. A reversible or equilibrium process is a process that can be reversed by infinitesimal changes in the conditions of the process and in which the system is always in, or very close to, thermal equilibrium. All other thermodynamic processes are irreversible.
Heat Engine	A heat engine takes heat Q_H from a source, converts part of the heat to work W, and discards the remaining heat $\left|Q_C\right|$ at a lower temperature. The heat engine's thermal efficiency e is $$e = \frac{W}{Q_H} = 1 + \frac{Q_C}{Q_H} = 1 - \frac{\left|Q_C\right|}{\left|Q_H\right|}.$$
Otto Cycle	A gasoline engine operating in the Otto cycle has a theoretical maximum thermal efficiency given by $$e = 1 - \frac{1}{r^{\gamma-1}},$$ where r is the compression ratio.
Refrigerator	A refrigerator takes heat Q_C from a cold source, performs work W, and discards the heat $\left|Q_H\right|$ to a warmer source. The performance coefficient K is $$K = \frac{Q_C}{W} = \frac{\left|Q_C\right|}{\left|Q_H\right| - \left|Q_C\right|}.$$
The Second Law of Thermodynamics	The second law of thermodynamics states that it is impossible for any cyclic system to convert heat completely into work. It also states that no cyclic process can transfer heat from a cold place to a hot place without any input of work.
Carnot Cycle	The Carnot cycle operates between two heat reservoirs and represents the most efficient heat engine. The Carnot cycle combines the reversible adiabatic and isothermal expansion and contraction between two heat reservoirs at temperatures T_H and T_C, respectively. The efficiency of the Carnot cycle is $$e_{\text{Carnot}} = 1 - \frac{T_C}{T_H} = \frac{T_H - T_C}{T_H}.$$
Entropy	Entropy is a quantitative measure of the disorder of a system. The entropy change in a reversible thermodynamic system is $$\Delta S = \int_1^2 \frac{dQ}{T}.$$ The second law of thermodynamics can be stated as "The entropy of an isolated system may increase, but not decrease. The total entropy of a system interacting with its surroundings may never decrease."

Conceptual Questions

1: Cleaning your room

Your parents are always nagging you about cleaning your room. After learning about the second law of thermodynamics, you explain to your parents that it is impossible to clean your room, since cleaning would reduce the entropy inside your room and violate the second law of thermodynamics. Your mother recalls her college physics course and convinces you that you can clean your room without violating the second law. How does she convince you?

Solution

IDENTIFY, SET UP, AND EXECUTE Your mother agrees with you that the entropy of a closed system can never decrease. But she notes that when you clean your room, the system consists of you plus your belongings. You can decrease the entropy of your belongings in your room by increasing the entropy of your body, as long as the total entropy increases. You can certainly clean your room!

EVALUATE This problem shows how the entropy of isolated components in a system may decrease as long as the system's total entropy increases. It also shows that you shouldn't argue with your mother, although you may want to try the argument on your father, who doesn't remember his physics course.

2: Leaving a refrigerator door open to cool a room

When the air-conditioning system at your house fails, your younger brother suggests leaving the refrigerator door open to cool the house. Is this method effective?

Solution

IDENTIFY, SET UP, AND EXECUTE A refrigerator cools its contents by taking heat away from the contents, performing work, and expelling heat to a warmer region. The heat expelled is always greater than the heat removed from the contents. The refrigerator must add net heat to its surroundings. Opening the refrigerator will result in a warmer room, so it is not an effective method of cooling the room.

EVALUATE Can opening the refrigerator warm the house on a cold day? Yes, since it must expel heat to operate. It wouldn't be a very efficient heat source, but it would provide some heat to the room.

3: Water as a fuel

Some people have suggested using water as a clean fuel. The idea is to break apart water molecules into hydrogen and oxygen. Then, when the hydrogen (combined with water) is burned, it produces energy without pollution. How does the second law of thermodynamics relate to this idea?

Solution

IDENTIFY, SET UP, AND EXECUTE Because the breaking apart of the water and the burning of hydrogen constitutes a reversible cycle, the net entropy must increase. The process may actually create *more* pollution, since it takes more energy to dissociate the water than is recovered by burning the hydrogen. For example, if gasoline is used to generate the hydrogen, it would take more gasoline to generate an equivalent amount of hydrogen-based power than if the gasoline were used to operate the vehicle directly.

EVALUATE There is the possibility that pollution would be reduced. If the hydrogen-generating plants installed high-quality pollution filters, than there could be less pollution generated overall by the plant compared with the pollution generated by many cars. However, a hydrogen-burning car will always require more total energy to operate. Hydrogen should probably be considered an alternative energy storage method.

Problems

1: Work in a heat engine

A heat engine carries 0.2 mol of argon through the cyclic process shown in Figure 20.1. Process ab is isochoric, process bc is adiabatic, and process ca is isobaric at a pressure of 2.0 atm. Find the net work done by the gas in the complete cycle. The temperatures at the of endpoints of the process are $T_a = 290$ K, $T_b = 650$ K, and $T_c = 440$K.

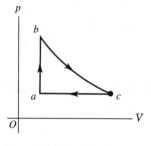

Figure 20.1 Problem 1.

Solution

IDENTIFY Use the principles of heat engines to find the work done.

SET UP We'll break the cycle up into processes and find the work done during each process. We'll need to find the pressure and volume at the three points before finding the work. Argon is an ideal gas, so we'll use the ideal-gas relations.

EXECUTE Starting at point a, we find the volume V_a from the ideal-gas equation:

$$V_a = \frac{nRT_a}{p_a} = \frac{(0.2 \text{ mol})(8.31 \text{ J/mol/K})(290 \text{ K})}{(2.02 \times 10^5 \text{ Pa})} = 2.39 \times 10^{-3} \text{ m}^3.$$

At point b, the volume is the same as at point a. We find the pressure at b:

$$p_b = \frac{nRT_b}{V_b} = \frac{(0.2 \text{ mol})(8.31 \text{ J/mol/K})(650 \text{ K})}{(2.39 \times 10^{-3} \text{ m}^3)} = 4.52 \times 10^5 \text{ Pa}.$$

We find the volume V_c at c:

$$V_c = \frac{nRT_c}{p_c} = \frac{(0.2 \text{ mol})(8.31 \text{ J/mol/K})(440 \text{ K})}{(2.02 \times 10^5 \text{ Pa})} = 3.62 \times 10^{-3} \text{ m}^3.$$

With these values, we find the work done during each process. There is no work done during the process ab, since it is an isochoric process. Process bc is adiabatic and there is no heat exchanged. The work is opposite the change in internal energy, or

$$W_{bc} = -\Delta U_{bc} = -nC_V(T_c - T_b) = -(0.2 \text{ mol})(12.47 \text{ J/mol/K})(440 \text{ K} - 650 \text{ K}) = 524 \text{ J},$$

where we used the molar heat capacity at constant volume for argon ($C_V = 12.47$ J/mol/K, from Table 19.1). Process ca is isobaric and the work is

$$W_{ca} = p_c \Delta V_{ca} = (2.02 \times 10^5 \text{ Pa})((2.39 \times 10^{-3} \text{ m}^3) - (3.62 \times 10^{-3} \text{ m}^3)) = -248 \text{ J}.$$

Note that the work done by the gas is negative, since it is compressed in ca. The work done during the complete cycle is the sum of the separate amounts of work done during each of the three processes:

$$W = W_{ab} + W_{bc} + W_{ca} = 0 + 524 \text{ J} - 248 \text{ J} = 276 \text{ J}.$$

The gas does 276 J of work in one cycle.

EVALUATE Since the area inside the pV cycle diagram is equal to the work, our positive result is in agreement with the area shown in the diagram. Much of this problem is based on what we learned in Chapter 19. We are now combining the processes of Chapter 19 into a complete cycle.

2: **Efficiency of a heat engine**

Find the thermal efficiency of an engine that operates in accordance with the cycle shown in Figure 20.2, in which 2 moles of helium stored at 2.0 atm in a 10-liter vessel starts at point a, undergoes an isochoric process to quadruple its pressure at point b, triples in volume in an isobaric expansion to point c, reduces its pressure to one-fourth its pressure at point c through an isochoric process at point d, and goes through an isobaric compression reducing its volume by one-third to return to point a.

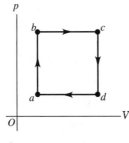

Figure 20.2 Problem 2.

Solution

IDENTIFY The target variable is the thermal efficiency of the engine.

SET UP We need to know the work and heat of the cycle to find the efficiency of the engine. We break the cycle up into processes and use our knowledge of isochoric (constant-volume) and isobaric (constant-pressure) processes. We are given the changes in pressure and volume, so we can proceed immediately to calculating the work done during each of the four processes. Helium is an ideal gas, so we can use the ideal-gas relations.

EXECUTE We start by determining the pressure and volume at the points b, c, and d. The pressure at points b and c is 8 atm ($4P_a$) and the pressure at d is 2 atm (P_a). The volume at b is 10 liters (V_a) and the volume at c and d is 30 liters ($3V_a$). Next, we find the work for each process. No work is done during processes ab and cd, since they are isochoric. Process bc is isobaric and the work done is

$$W_{bc} = p_b \Delta V_{bc} = (8.08 \times 10^5 \, \text{Pa})((30 \times 10^{-3} \, \text{m}^3) - (10 \times 10^{-3} \, \text{m}^3)) = 16{,}200 \, \text{J}.$$

Process da is also isobaric and the work done is

$$W_{da} = p_d \Delta V_{da} = (2.02 \times 10^5 \, \text{Pa})((10 \times 10^{-3} \, \text{m}^3) - (30 \times 10^{-3} \, \text{m}^3)) = -4040 \, \text{J}.$$

The total work is the sum of the separate amounts of work done during each of the four processes:

$$W = W_{ab} + W_{bc} + W_{cd} + W_{da} = 0 + 16{,}200 \, \text{J} + 0 - 4040 \, \text{J} = 12{,}100 \, \text{J}.$$

We need to find the heat flowing into the engine. Heat flows into the engine during processes ab and bc. To find the heat flow into the engine, we need to know the temperature at points a, b, and c. The ideal-gas equation gives

$$T_a = \frac{p_a V_a}{nR} = \frac{(2.02 \times 10^5 \, \text{Pa})(10 \times 10^{-3} \, \text{m}^3)}{(2 \, \text{mol})(8.31 \, \text{J/mol/K})} = 122 \, \text{K},$$

$$T_b = \frac{p_b V_b}{nR} = \frac{(8.08 \times 10^5 \, \text{Pa})(10 \times 10^{-3} \, \text{m}^3)}{(2 \, \text{mol})(8.31 \, \text{J/mol/K})} = 486 \, \text{K},$$

$$T_c = \frac{p_c V_c}{nR} = \frac{(8.08 \times 10^5 \, \text{Pa})(30 \times 10^{-3} \, \text{m}^3)}{(2 \, \text{mol})(8.31 \, \text{J/mol/K})} = 1460 \, \text{K}.$$

The heat flow in process ab (carried out at constant volume) is

$$Q_{ab} = nC_V \Delta T_{ab} = (2 \, \text{mol})(12.47 \, \text{J/mol/K})(486 \, \text{K} - 122 \, \text{K}) = 9080 \, \text{J},$$

where we used the molar heat capacity at constant volume for helium ($C_V = 12.47 \, \text{J/mol/K}$, from Table 19.1). The heat flow in process bc (carried out at constant pressure) is

$$Q_{bc} = nC_P \Delta T_{bc} = (2 \, \text{mol})(20.78 \, \text{J/mol/K})(1460 \, \text{K} - 486 \, \text{K}) = 40{,}500 \, \text{J},$$

where we used the molar heat capacity at constant pressure for helium ($C_P = 20.78 \, \text{J/mol/K}$, from Table 19.1). The total heat flowing into the engine in one cycle is therefore

$$Q_H = Q_{ab} + Q_{bc} = 40{,}500 \, \text{J} + 9080 \, \text{J} = 49{,}600 \, \text{J}.$$

The efficiency of the engine is

$$e = \frac{W}{Q_H} = \frac{12{,}100 \, \text{J}}{49{,}600 \, \text{J}} = 24.4\%.$$

EVALUATE We see that the engine is 24.4% efficient. Note that, to find the efficiency, we started by determining the state variables for the points on the pV diagram. Then we found the work and heat flow in the cycle and combined these two pieces of information to solve the problem.

3: Efficiency of a diesel engine

Find the thermal efficiency of the diesel cycle shown in Figure 20.3 for an engine having a compression ratio $R = 18$, an expansion ratio $E = 6 = RV/V_C$, and $C_P/C_V = 1.4$.

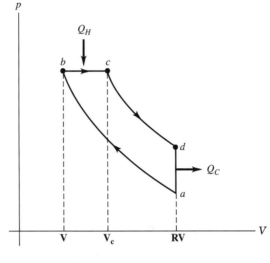

Figure 20.3 Problem 3.

Solution

IDENTIFY The target variable is the thermal efficiency of the engine. The efficiency involves the heat into and out of the system.

SET UP We start by finding the heat into and out of the system. Along the adiabatic lines, no heat is transferred. We will be able to find the heat in terms of the changes in temperature and then solve for the temperatures. We will substitute the compression and expansion factors where appropriate.

EXECUTE The efficiency is given by

$$e = 1 + \frac{Q_C}{Q_H}.$$

The heat entering the cycle between points b and c is given by

$$Q_{bc} = nC_p(T_c - T_b).$$

The heat leaving the cycle between points d and a is given by

$$Q_{da} = nC_V(T_a - T_d).$$

No heat is exchanged between any other pairs of points, as they lie upon adiabatic lines. The efficiency is then

$$e = 1 + \frac{T_a - T_d}{\gamma(T_c - T_b)},$$

where we replaced the ratio of the specific heats with γ. To find the efficiency, we need to find the temperatures. We seek relations between the temperatures that should cancel out. Points b and c are at the same pressure, so we have

$$\frac{V}{T_b} = \frac{V_c}{T_c}.$$

The ratio of the volumes can be substituted for the volumes:

$$T_c = T_b \frac{R}{E}.$$

To relate these temperatures to the temperature at a, we use the adiabatic relation

$$T_b V^{\gamma-1} = T_a V_a^{\gamma-1} = T_a (RV)^{\gamma-1}.$$

So we have

$$T_b = T_a R^{\gamma-1}, \qquad T_c = T_a R^{\gamma-1} \frac{R}{E}.$$

We find the temperature at point d by using the other adiabatic line:

$$T_c V_c^{\gamma-1} = T_d V_d^{\gamma-1} = T_d (RV)^{\gamma-1}.$$

Solving for T_d yields

$$T_d = \left(\frac{R}{E}\right)^\gamma T_a.$$

We now replace the temperatures with their expressions in terms of T_a to find the efficiency:

$$e = 1 + \frac{T_a - T_d}{\gamma(T_c - T_b)}$$

$$= 1 + \frac{(1 - (R/E)^\gamma)}{\gamma(R^\gamma/E - R^{\gamma-1})}$$

$$= 1 + \frac{(1 - (18/6)^{1.4})}{1.4(18^{1.4}/6 - 18^{0.4})} = 0.59.$$

The efficiency is 59%.

EVALUATE We see that the diesel engine is 59% efficient. Diesel engines are generally more efficient than gasoline engines.

4: Entropy change in melting ice

A heat reservoir at 50°C is used to melt 25 kg of ice at 0°C. What is the entropy change in the melted ice? What is the entropy change in the reservoir? What is the total entropy change in the system?

Solution

IDENTIFY The target variables are the entropy changes for the ice, reservoir, and total system.

SET UP Entropy change in a reversible process is equal to the heat transferred divided by the temperature of the material. We can find the heat required to melt the ice, which must be equal to the heat provided by the reservoir.

EXECUTE The heat required to melt the ice is given by the heat of fusion relation and is

$$Q = mL_f = (25 \text{ kg})(335 \times 10^3 \text{ J/kg}) = 8.375 \times 10^6 \text{ J},$$

where we used the latent heat of fusion for ice $(335 \times 10^3 \text{ J/kg})$. The change in entropy for the water is

$$\Delta S_{ice} = \frac{Q}{T} = \frac{8.375 \times 10^6 \text{ J}}{273 \text{ K}} = 30{,}700 \text{ J/K}.$$

The change in entropy for the reservoir is equal to the heat leaving the reservoir divided by the temperature of the reservoir. The reservoir loses as much heat as the ice gains. The entropy change is

$$\Delta S_{reservoir} = \frac{-Q}{T} = \frac{-8.375 \times 10^6 \text{ J}}{323 \text{ K}} = -25{,}900 \text{ J/K}.$$

The total change in entropy is the sum of the entropies for the ice and reservoir:

$$\Delta S_{total} = \Delta S_{ice} + \Delta S_{reservoir} = 30{,}700 \text{ J/K} - 25{,}900 \text{ J/K} = 4800 \text{ J/K}.$$

The entropy of the ice increases by 30,700 J/K, the entropy of the reservoir decreases by 25,900 J/K, and the total entropy of the system increases by 4800 J/K.

EVALUATE The entropy of the reservoir decreased, but this does not violate the second law, since the system is the combination of the ice, the water, and the reservoir. The total change in entropy is positive, as expected.

5: Entropy change in isothermal expansion

Find the change in entropy for an ideal gas that undergoes an isothermal expansion from an initial volume to a final volume that is twice the initial volume.

Solution

IDENTIFY The target variable is the entropy change for the gas.

SET UP Entropy change is related to the work and the temperature of the material. For an ideal gas, the internal energy doesn't change. We'll use the standard definitions to find the entropy change in the process.

EXECUTE Writing the change in entropy in terms of work and temperature gives

$$TdS = dQ = 0 + dW,$$

since the change in internal energy is zero. The work done by an ideal gas undergoing an isothermal process is given by

$$dW = pdV = \frac{nRT}{V}dV.$$

Combining the two equations yields an expression for the change in entropy:

$$dS = \frac{nR}{V}dV.$$

We integrate to find the change:

$$S = \int_{V_1}^{V_2} \frac{nR}{V} dV = nR \ln V \Big|_V^{2V} = nR \ln 2.$$

The change in entropy is $R \ln 2$ for each mole of gas.

EVALUATE The entropy of the gas increased, as expected.

Try It Yourself!

1: Efficiency of a heat engine

Following the cycle shown in Figure 20.4, find the efficiency of a heat engine using an ideal monatomic gas as its working substance. Take $\gamma = 1.4$ and $r = V_b/V_a = 2.5$.

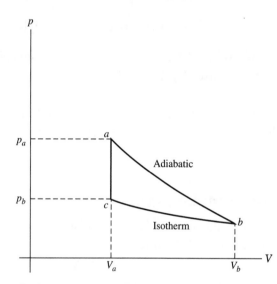

Figure 20.4 Try It Yourself 1.

Solution Checkpoints

IDENTIFY AND SET UP Break the process up into segments and the heat exchanged during each segment.

EXECUTE No heat is exchanged in segment ab. (Why?) Along segment bc, heat is removed from the system and is equivalent to

$$Q = nRT_c \ln(V_a/V_b).$$

Along segment ca, heat is put into the system and is equivalent to

$$Q = nC_V(T_a - T_c).$$

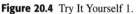

The temperatures are written in terms of T_c to find the efficiency, which is

$$e = 1 - \frac{((\gamma - 1) \ln r)}{r^{\gamma-1} - 1}.$$

EVALUATE What effect does increasing r have on the efficiency?

2: Entropy in mixed water

Equal volumes of water at 80°C and 20°C are mixed together. Find the increase in entropy for a total volume of 1.0 m³.

Solution Checkpoints

IDENTIFY AND SET UP The final temperature of the mixture should be halfway between the initial temperatures of the two volumes of water if the heat capacities are constant and independent of temperature. No process is noted, so we can choose any reversible process. Assume that the process keeps the volume constant.

EXECUTE The change in entropy for the system is

$$\Delta S = \int_1^2 \frac{dQ}{T} = C_V \int_{T_1}^{T_2} \frac{dT}{T} = C_V \left(\frac{T_2}{T_1} \right),$$

where the temperatures are in kelvins. The entropy of the hot water decreases, while the entropy of the cold water increases. The total change in entropy is 3.62×10^4 J/K.

EVALUATE How did you determine C_V?